COMPILATION, CRITICAL EVALUATION, AND DISTRIBUTION OF STELLAR DATA

ASTROPHYSICS AND
SPACE SCIENCE LIBRARY

A SERIES OF BOOKS ON THE RECENT DEVELOPMENTS

OF SPACE SCIENCE AND OF GENERAL GEOPHYSICS AND ASTROPHYSICS

PUBLISHED IN CONNECTION WITH THE JOURNAL

SPACE SCIENCE REVIEWS

VOLUME 64

PROCEEDINGS

COMPILATION, CRITICAL EVALUATION AND DISTRIBUTION OF STELLAR DATA

PROCEEDINGS OF THE INTERNATIONAL
ASTRONOMICAL UNION COLLOQUIUM No. 35,
HELD AT STRASBOURG, FRANCE,
19-21 AUGUST, 1976

Edited by

C. JASCHEK
Observatoire de Strasbourg, France

and

G. A. WILKINS
Royal Greenwich Observatory, U.K.

D. REIDEL PUBLISHING COMPANY

DORDRECHT-HOLLAND/BOSTON-U.S.A.

ISBN-13: 978-94-010-1216-4   e-ISNB-13: 978-94-010-1214-0

DOI: 10.1007/978-94-010-1214-0

Published by D. Reidel Publishing Company,
P.O. Box 17, Dordrecht, Holland

Sold and distributed in the U.S.A., Canada and Mexico
by D. Reidel Publishing Company, Inc.
Lincoln Building, 160 Old Derby Street, Hingham,
Mass. 02043, U.S.A.

TABLE OF CONTENTS

Introduction, by the Editors   ix

Organizing Committees   xi

List of Participants   xiii

PART I.   STANDARDS FOR THE PRESENTATION OF DATA

Chairman, Dr. P. Lacroute

1. M.S. Davis   (Invited paper) Standards, management
                       and security of astronomical data sets   3

2. J.C. Mermilliod   Principles of a coded numbering system
                       and its application to open clusters   19

3. H. Eichhorn   Thoughts on the form of presentation of
                       star catalogue data   23

4. M.J. Collins   A bibliography of astronomical catalogues
                       and the organization of stellar
                       designations   25

5. F. Ochsenbein,   The catalogue of stellar identifications   31
   D. Egret and
   M. Bischoff

6. W. Gliese   Desiderata for the catalogue of nearby
                       stars   37

7. F. Spite   A proposal for non-ambiguous designation
                       of stars and stellar objects   41

8. G.A. Wilkins   The use of SI units in astronomy   43

PART II.   ACQUISITION AND PROCESSING TECHNIQUES

Chairman, Dr. D. Hoffleit

9. G. Westerhout   (Invited paper) The influence of acquisi-
                       tion techniques on the compilation of
                       astronomical data   49

10. I. Mistrik   Data processing and analysis for space-
                        based astronomy   61

11. R.H. Harten   The use of standardized data formats
    and   with the Westerbork radio telescope   69
    T.A.Th. Spoelstra

12. H.G. Walter   Data storage requirements in relation
                        to radio interferometry observations   77

13. K. Nandy   Ultraviolet photometry from the S2/68
                        observations in the TD1 satellite   85

14. S.B. Parsons   The University of Texas catalog of
    and   ultraviolet and optical stellar data   93
    J.D. Wray

            PART III. THE CRITICAL EVALUATION OF DATA
                   Chairman, Dr. G.A. Wilkins

15. A.B. Underhill,   (Invited paper) The critical evaluation
    J.M. Mead and   of stellar data   105
    T.A. Nagy

16. B. Nicolet and   Critical evaluation of photometric data   121
    B. Hauck

17. W.P. Bidelman   Some thoughts on astronomical data files   125

18. B.V. Kukarkin,   Work on the general catalogue of
    P.N. Kholopov   variable stars   131
    and N.N. Kireeva

19. A.N. Argue and   A catalogue of photometric sequences   135
    E.W. Miller   Supplement no. 2

20. A.H. Batten   Catalogues of spectroscopic binaries   141

21. F.B. Wood   The Princeton-Pennsylvania-Florida card
                        catalogue of eclipsing variables   147

22. D. Hoffleit   Plans for a new edition of the bright
    and   star catalogue   151
    C. Jaschek

            PART IV. THE DISTRIBUTION OF DATA
                 Chairman, Dr. R.L. Duncombe

23. B. Hauck   (Invited paper) Data distribution   155

24. J. Mead and   Retrieval techniques and graphics
    T.A. Nagy   displays using a computerized stellar
                        data base   161

25. R.S. Dixon   A master list of non-stellar objects   167

26. M.C. Lortet   Proposal for a data centre on
                        galactic non-stellar objects   171

27. F. Ochsenbein   Main features of the stellar biblio-
    and F. Spite   graphic file   175

28. C.E. Worley   The visual double star catalogues   179

29. W.F. van   Systematic differences in trigonometric
    Altena,   parallaxes from different observatories   183
    E.D. Hoffleit
    and H.A. Smith

30. Y. Terashita   The astronomical data systems group in
                        Japan   191

31. D.D.   Stellar data and computing facilities
    Polojentsev   at the Pulkovo Observatory   195

32. A.D. Fiala and   Astronomical data files at the U.S.
    P.K. Seidelmann   Naval Observatory: star catalogues,
                        ephemerides, and observations   199

PART V. EXISTING FACILITIES AND FUTURE ROLE OF DATA CENTRES

Chairman, Dr. G.A. Wilkins

33. C. Jaschek   (Invited paper) Survey of existing
                        facilities   205

34. W.G. Baum   Existing data centers and their
                        future role   281

35. W. Buscombe   Third general catalogue of MK spectral
                        classifications   285

36. G.A. Wilkins   The future role of data centres in
                        astronomy   287

37. W.D. Heintz   Astronomical data coordination: a
                        perpetual task   289

38. R.H. Garstang   Concluding remarks   293

Report of Discussions, by G.A. Wilkins   295

INTRODUCTION

The principal purpose of IAU Colloquium No. 35 was to discuss those aspects of the techniques of the compilation, evaluation, and distribution of data that are common to astrometry, photometry and spectrometry of stars and stellar systems. In the announcement of the Colloquium, it was suggested that there would be special emphasis on the techniques of quality control, and on the standards for the presentation of numerical data in both printed and computer-readable form. As the meeting progressed it became clear that the lack of a standard, unambiguous system for the identification of stellar objects was a source of much confusion and inefficiency in the use of existing data files. This and other such matters were the subject of further discussions by Commission 5 at the General Assembly of the International Astronomical Union (IAU), which was held at Grenoble during the following fortnight, 24 August - 2 September 1976.

The proposal for the Colloquium was prepared by J. Jung, who was then Director of the Centre des Données Stellaires at Strasbourg, and G.A. Wilkins, Chairman of the IAU Working Group on Numerical Data, and was accepted by the IAU Executive Committee on the recommendation of the President of Commission 5, with the support of Commissions 25, 29 and 45. The Scientific Organising Committee consisted of W. Fricke, B. Hauck, C. Jaschek, J. Jung, B. Kukarkin, P. Lacroute, A. Underhill and G.A. Wilkins (Chairman). The Local Organising Committee consisted of A. Florsch, C. Jaschek, P. Lacroute (Chairman), and F. Ochsenbein. C. Jaschek acted as the Secretary of the Colloquium.

The Colloquium was held at the kind invitation of the Université Louis Pasteur at Strasbourg. Generous financial assistance was given by the National French Committee for Astronomy, the U.E.R. de Mathématiques (University of Strasbourg), the Conseil Regional de la region d'Alsace, the Municipalité de la Ville de Strasbourg and the Strasbourg Observatory, and some travel grants were given by the International Astronomical Union.

An informal gathering of the participants was held at the Observatory of Strasbourg during the evening of Wednesday, 18 August. The Colloquium opened at 0900 a.m. on Thursday, 19 August, when Professor P. Lacroute invited Professor P. Karli, President of the University, to speak; he welcomed the participants and gave a brief account of the history and relationships of the three universities at Strasbourg. G.A. Wilkins replied on behalf of the participants; he then reviewed the aims of the Colloquium and described the arrangement of the scientific programme. Then Dr. A. Florsch, who had been recently appointed Director of the Strasbourg Observatory in succession to Professor P. Lacroute, welcomed the participants on behalf of the Observatory and described the local arrangements and the social functions which would be held in the evenings.

The programme consisted of five sessions, each of which was opened by the presentation of an invited paper on a topic of general interest. After the general discussions on these papers there were short presentations of contributed papers, which often led to further discussion. Whenever possible the contributed papers were arranged so that their topics were related to that of the preceding invited paper, but for various reasons this could not always be done. The programme is given on pages v-vii.

The papers formally presented at the Colloquium are given in full, or by abstract. The discussions have been summarised (see pages 295-316) and, where appropriate, the sequence has been changed to bring together related points. Use has been made of the notes which were provided by the participants after the discussions, and which have been transcribed under the supervision of C. Jaschek, as well as of other manuscript notes and tape recordings. It is hoped that no-one's views have been seriously misrepresented, and that no significant comments have been omitted, but it has not been practicable to verify the correctness of this report.

At the conclusion of the Colloquium, Professor W.D. Heintz, Vice-President of IAU Commission 5, expressed the thanks of the participants to the organisers and especially to the staff of the Observatory at Strasbourg for making all the necessary arrangements for what had proved to be a very interesting and useful meeting.

It is an agreeable pleasure to thank Misses Ch. Bruneau and S. Miller, who prepared the manuscript of the proceedings for printing.

The Editors

ORGANIZING COMMITTEES OF THE COLLOQUIUM

SCIENTIFIC ORGANIZING COMMITTEE

G. A. Wilkins (Chairman)
W. Fricke
B. Hauck
C. Jaschek (Secretary)
J. Jung
B. V. Kukarkin
P. Lacroute
A. B. Underhill

LOCAL ORGANIZING COMMITTEE

P. Lacroute (Chairman)
A. Florsch
C. Jaschek
F. Ochsenbein

LIST OF PARTICIPANTS

AUSTRIA

    Eichhorn H.K.

BELGIUM

    Heck A.

BULGARIA

    Kalinkov M.

CANADA

    Batten A.M.

FRANCE

    Acker A.
    Barbier M.
    Bischoff M.
    Bru P.
    Delhaye J.
    Dubois P.
    Egret D.
    Florsch A.
    Fresneau A.
    Gomez A.
    Jaschek C.
    Jaschek M.
    Lacroute P.
    Lortet M.C.
    Mennessier M.O.

    Muller P.
    Ochsenbein F.
    Spite F.
    Valbousquet A.
    Wenger M.

GERMANY, FEDERAL REPUBLIC

    Du Mont B.
    Gliese W.
    Jahreiss H.
    Mistrik I.
    Walter H.G.

JAPAN

    Terashita Y.

NETHERLANDS

    Spoelstra J.A.
    Van Herk G.

SOUTH AFRICA

    Hers J.

SWEDEN

    Lynga G.
    Oja T.

SWITZERLAND

    Hauck B.
    Mermilliod J.Cl.
    Morel M.
    Nicolet B.

UNITED KINGDOM

    Argue A.N.
    Collins M.J.
    Nandy K.
    Wilkins G.A.
    Yallop B.D.

UNION OF SOVIET SOCIALIST REPUBLICS

    Abalakine V.K.
    Straizys V.

UNITED STATES OF AMERICA

    Baum W.A.
    Bidelman W.P.
    Davis M.S.
    Davis R.J.
    Dixon R.S.
    Duncombe R.L.
    Garstang R.H.
    Heintz W.D.
    Hoffleit D.
    Luyten W.
    Mead J.
    Parsons S.B.
    Slettebak A.
    Underhill A.B.
    Van Altena W.F.
    Westerhout G.
    Wing R.F.
    Wood F.B.
    Worley C.E.

VATICAN CITY STATE

    McCarthy M.F.

PART I

STANDARDS FOR THE PRESENTATION OF DATA

STANDARDS, MANAGEMENT AND SECURITY OF ASTRONOMICAL DATA SETS

M. S. Davis

Department of Physics and Astronomy
The University of North Carolina at Chapel Hill

INTRODUCTION

Astronomers have historically been among the first to apply or develop, at times, new technologies in the furtherance of their science. This has been especially true in the use of computers from their archaic, antediluvian beginnings to the present highly-developed, time-sharing, multiprocessing, teleprocessing systems. Thus, among the earliest applications in the last half century was the use by L. J. Comrie at the Greenwich Observatory of Hollerith machines for the construction of tables (Comrie, 1928). In 1940 Eckert described punched card methods for numerical integration, computation of a numerical lunar theory, computation of planetary perturbations, as well as applications in photometry and construction of star catalogues (Eckert, 1940). Indeed, one of the earliest collections of files for general use was available at the Watson Astronomical Bureau and included Boss' General Catalogue of 33,342 stars, A. G. Catalogues, Yale Zone Catalogues, and Kohlschütter's Catalogue.

No doubt the Space Age, ushered in by the first Soviet sputnik in 1957 along with the evolution of computers occurring then, heightened the need for extensive data files, at least in some areas. For example, the Yale Catalogues were used as a data bank in computer programs which reduced observations of artificial satellites necessitated by the extremely large number of observations on a growing number of satellites, and required on short time scales for orbit correction.

This Colloquium is testimony to the growth and spread of astronomical data files in recent years to virtually every field

C. Jaschek and G. A. Wilkins (eds.), Compilation, Critical Evaluation, and Distribution of Stellar Data. 3-17.
Copyright © 1977 by D. Reidel Publishing Company, Dordrecht-Holland. All Rights Reserved.

of astronomy from the earth, to the solar system, to stellar data,
to the galaxy, to external galaxies, including reference material
like tabulations of atomic spectral data, ephemerides, optical,
radio, and x-ray observations. The use of these materials in a
variety of fields has more and more pointedly revealed problems
arising from data sets, most of which have been designed for their
own special purposes, without necessarily considering the possible
use of this information for other purposes, or, indeed, as part
of a repository in large, astronomical, data banks for general use
in a large, possible variety of ways.

It is the recognition of a large constellation of problems
involving data sets, focussing attention on Compilation, Critical
Evaluations and Distribution of Stellar Data that is the raison
d'être for this Colloquium of the I.A.U.

ORGANIZATIONS DEVOTED TO DATA BASE MANAGEMENT

Astronomers should be aware of at least three organizations
devoted to the subject of Data Bases and their management:

1. CODASYL (Conference on Data Systems Languages) was organ-
    ized in 1959. Thanks to a number of Task Groups, in
    particular, the DBTG (Data Base Task Group), CODASYL
    has produced a number of consequential reports concern-
    ing all aspects of Data Bases and their management in a
    variety of milieux (CODASYL 1969, 1971), and there exists
    today a staggering literature on the subject. The im-
    plementations of the CODASYL reports have used the high-
    level, procedural language COBOL and, hence, may be a
    disadvantage from the astronomer's point of view. None-
    theless, the fundamental concepts developed are sound
    and based upon vast experience over an extremely wide
    spectrum of applications.

2. GUIDE-SHARE Data Base Requirements Group (GUIDE-SHARE
    1970). This group has approached the subject without
    defining the syntax of languages to be used and has been
    primarily concerned with developing concepts of impor-
    tance. There is not surprisingly a great similarity in
    the ideas and principles developed by both the CODASYL
    and the GUIDE-SHARE groups. The group making the GUIDE-
    SHARE recommendations came from 40 diverse organizations
    representing years of experience in such areas as bank-
    ing, life insurance, machine manufacture, government
    agencies, and universities.

3. CODATA (Committee on Data for Science and Technology,
    founded in 1968), is an organ of the International

Council of Scientific Unions of which the I.A.U. is a
member and which is represented in it. Hence, it is of
especial interest to astronomers. (CODATA, 1968-1976)

While, for the most part, the vast literature dealing with Data
Bases and their management is largely irrelevant to the problems
of interest to astronomers today, much of the experience and
many of the ideas are, indeed, invaluable and will be even more
so as astronomical data bases are consolidated into data banks
which are to be used in sophisticated ways in the future.

DATA BASE MANAGEMENT SYSTEMS

Let us begin with a general Data Base Management System.
Figure 1 is a flow chart and description of such a system. For
the system to be viable, the Data Base must be reliable and main-
tainable in the sense that it can be updated and errors in it can
be corrected. Furthermore, it must be capable of search processes
to satisfy inquiries and requests made by users.

The Data Base Administrator is a person who defines important
parameters concerning the System, among which are:

1. The schema. A complete description of the Data Base.

2. Subschemas. Subsets of the schema made available to
    various users.

3. Security. In the present context security is meant to
    be protection of the Data Base against un-
    authorized access and changes in data. The
    concept of security has other connotations
    which will be elaborated on soon.

The Data Base Manager is a program used (1) by the Data Administrator to enforce his policies and (2) by users to access
and manipulate data. At this point let us more precisely define
several concepts:

1. Privacy. Protection against unauthorized access of data.

2. Integrity. Protection against corruption of data.

3. Security. In general, the term shares meanings between
    privacy and integrity, but most often is
    taken to be the same as privacy.

Security is generally enforced by operating techniques on the file
level during the running of a program and is most often accomplished

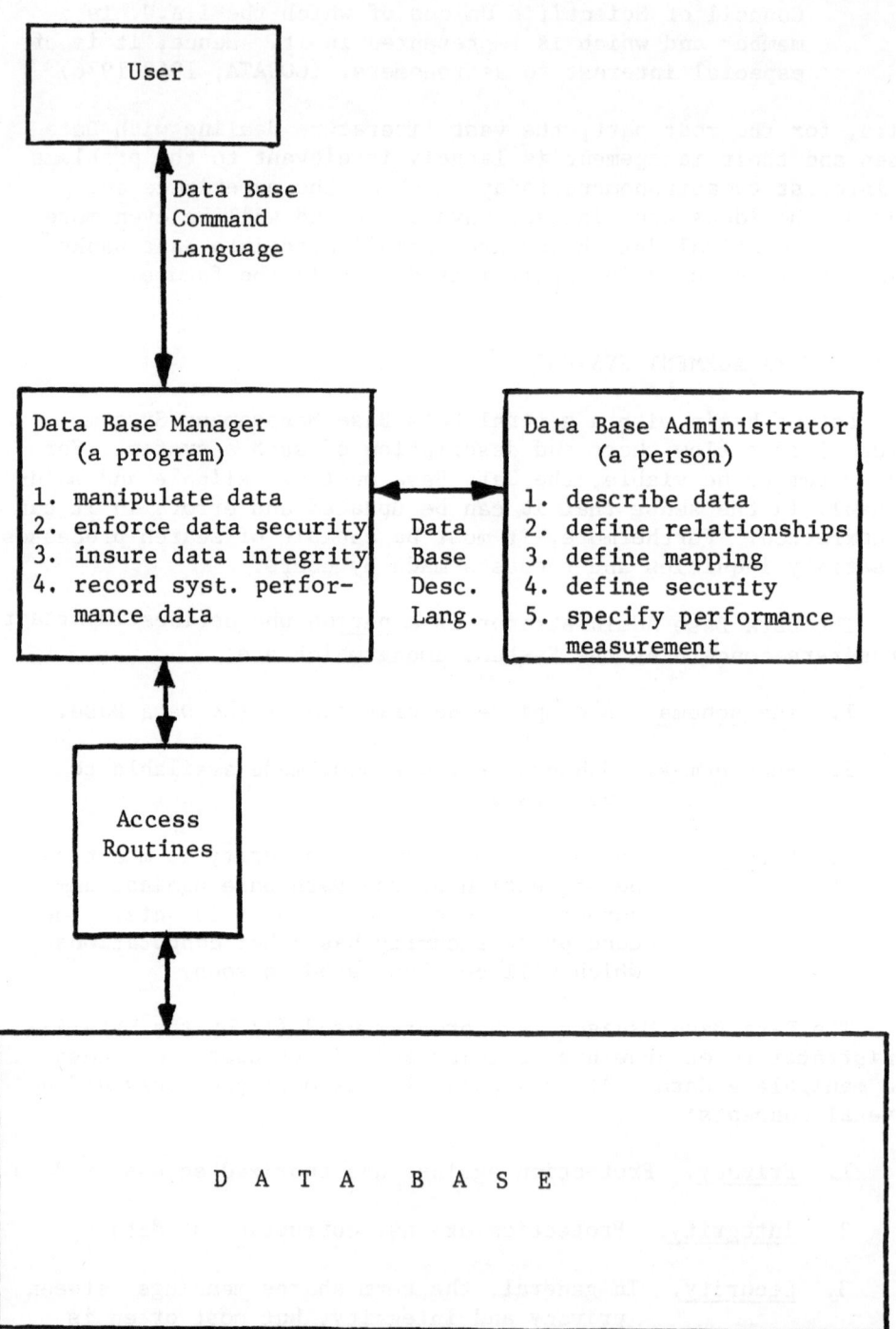

Figure 1. A General Data Base Management System

by making available <u>passwords</u> or <u>privacy keys</u> to <u>privacy locks</u> which are specified in the schema.

Another requirement for viability of the System is <u>Resilience</u> which may be defined as the capacity of a system to recover from errors of systems, program, or hardware types. Of some interest is the re-creation of a Data Base, or parts of it, after corruption or destruction has taken place. If the material in the Data Base is highly volatile (generally, not the case for most astronomical files, but which may apply to some during their creation), one common method for providing backup is periodic dumping of the files, or <u>journalization</u>, which is the logging of data transfers. One of the safest techniques when a file is to be updated, is to generate the new updated file called the "son" file. When this file is deemed correct, the "son" file replaces a previously saved file called the "father" file, the "father" file replaces the "grandfather" file and the "grandfather" file is destroyed. The following diagram outlines the process:

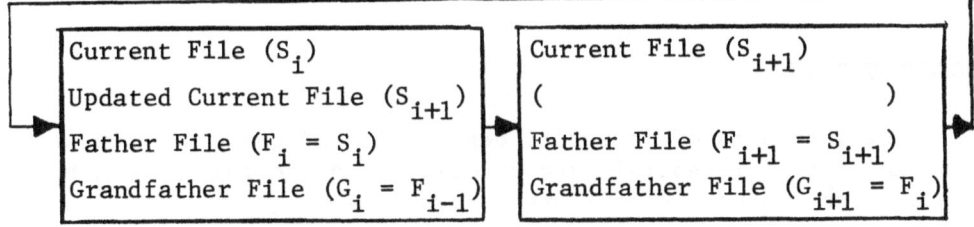

This strategy keeps a copy of the current file (the father file) and the previously updated copy, if it is necessary to reconstruct the current file, or if it is desirable to reconstruct a previous copy of the file. Clearly, older generations can be saved if an extremely high level of safety is needed.

CURRENT ASTRONOMICAL DATA BASES

Most astronomical data presently exist as individual files and reside in a variety of institutions. In 1970 the I.A.U. in collaboration with COSPAR, the Bureau des Longitudes and other scientific institutions established the International Information Bureau on Astronomical Ephemerides (B.I.I.E.A.) whose purpose is to provide information to the international scientific community on availability of astronomical ephemerides, star catalogues, etc. in machine-readable form for use in astronomical and space research. To date 125 information cards have been distributed (BIIEA 1971-1975).

In the context of a large Data Base System with concurrent processing, maintaining integrity is a complex matter involving, as indicated, privacy locks which may be applied to areas,

particularly data sets, records or even items. Furthermore, the
extent of such control has implications for efficiency, turn-
around and costs.

INTEGRITY OF ASTRONOMICAL DATA FILES

For the most part the astronomical files we are concerned
with are independent of large systems, and hence the integrity
problem is considerably simpler. Integrity for us is related
to the more fundamental questions of the existence and elimination of errors. In a larger sense integrity is also concerned
with information which is inclusive enough to make its use meaningful. For example, in a program to reduce observations of
artificial satellites, information may be needed concerning the
accuracy of the particular catalogue, or a statistical study em-
ploying the Yale Parallax Catalogue might need to employ probable
errors. These are among the considerations involved in the
"critical evaluation of stellar data."

DOCUMENTATION

The insurance of integrity begins with adequate documentation.
Documentation for general data description or definition can be
prepared on a number of different levels as part of:

1. the sources of the originally compiled data and possibly
    related procedure manuals.

2. a computer program which processes the data.

3. the computer files.

4. a computer program which manages the file.

5. a printout program.

These can be implemented in a number of ways (London, 1974).

Again, since most astronomical files are standalone and of
a historical nature in that they represent parameters measured
or calculated at particular times, the most appropriate documentation is that which makes the schema a part of the data files
themselves. I would urge general adoption of this idea for all
astronomical files. It is an extension of the self-documentation
which all well-documented programs have built into themselves.
The schema would be written in any natural language approved by
the I.A.U. and would not only describe each data item but would
provide information on accuracy, errors, caveats, equations used,

in short, the kind of information contained in the introduction
to catalogues. This would make each file totally independent
in the sense that it completely reproduces the original catalogue.
Preferably, I.A.U. conventions should be adhered to in notation
and usage.

These goals may not be attainable for many historical files
of astronomy but fortunately most of the files of interest to us
are already in a quite satisfactory format for uses and applications. The problems I have alluded to earlier will become real
to the designer of large astronomical Data Base Management Systems. In fact most users will design their own standalone programs to process the data they require from the files.

ERROR CORRECTION

Thus, the outstanding problem of integrity is the correctness
with which the machine-readable media have captured the original
source material. A second-order reliability on the correctness
of the files, is correction of data in the files discovered to be
fallacious in the source material. The first-order of correctness
involves the process of "proofreading" in some form or other.
Techniques in common use are:

1. proofreading of printouts (for greater safety at least
   one person to read the manuscript and at least one to
   read the printout),

2. comparing two or more files prepared independently from
   the same source material on the same or different media,

3. checking consecutivity of appropriate fields,

4. checking alphabetic or collating sequence order of
   appropriate fields,

5. checking for blanks, zeroes, etc.,

6. parameter ranges,

7. relational tests,

8. calculation (of calculable quantities) and comparison
   (as, for example, precession in star catalogues),

9. validity tests,

10. differencing, summation tests.

Some second-order corrections may be found by the above methods.
Others are often determined by the original compiler during revisions, updating, and other accesses to the file. Still other
corrections are discovered by users during applications when unexpected values or residuals appear. Whenever such errors are
discovered it is essential that the user communicate them to the
original source and to the repository whence he received the file.

As an example, the Astronomisches Rechen-Institut (ARI) has
its machine-readable files in agreement with the printed catalogues unless otherwise stated. Errors which are tabulated in
published errata lists have been corrected, as well as those discovered by the ARI and those communicated to it. These errors
are published from time to time and the ARI requests users of its
files to notify it if additional errors should be found.

The importance of data as error-free as possible, as well as
discussions of systematic and random errors can hardly be over-
emphasized. Macdonald raised the question just a few years ago
as to whether most data are worth owning (Macdonald, 1972).
While chairman of the Numerical Data Advisory Board of the National ·Research Council of the U.S.A. he was forced to conclude "No"
across the entire spectrum of research to the question posed.
His principal reasons were the lack of knowledge about the trust-
worthiness of the data and often a lack of trust of the measures
of uncertainty themselves. Fortunately, astronomers have a long
history and tradition of painstaking attention to such matters
in most areas of fundamental astronomy and if there is any concern it should be directed at maintenance of high standards of
error analysis in the newer disciplines.

STANDARDS FOR ASTRONOMICAL DATA FILES

My remarks so far have dealt primarily with astronomical
files of a historical nature comprising most of those currently
in existence as ascertained from the list compiled by Wilkins
in the Working Group on Numerical Data of I.A.U. Commission 5
and distributed to participants in Colloquium No. 35 or from the
cards of the BIIEA. A very few of the files, indeed, are volatile
in the sense that data items change, but when they do, it is
usually at a very slow rate. Examples of each kind are:

Historical Files   Volatile Files

Astrometric Star Catalogues   Observatories–staff and instruments
Minor Planets   Transition Probabilities

Some of these files are updated with more accurate determinations
(such as atomic energy levels, transition probabilities, or

planetary data), or additions (such as double stars, comets, x-ray sources or galaxies).

For this generation of astronomical files then, what should the standards be for compilers of these files? Summarizing what has been said, the standards should be as follows:

1. adherence to I.A.U. notation and conventions,

2. first-order correctness of machine-readable information as the result of scrupulous proofreading of material to ensure precise replication of source material,

3. second-order correctness of the information as the result of a variety of checks on the data itself to discover errors which may exist in the original catalogues or source data,

4. a schema which should be made a part of the file itself to make it totally independent and self-sufficient. The schema should have not only a complete description of data items but should contain all useful information including a discussion of errors, formula used, caveats as to use of the data, discussion of error-correcting techniques used and reliability of the file,

5. if applicable, tests and worked examples should be provided which may be used on the files,

6. periodic, or occasional, publication of errata or corrections of any type, or updating of files, informing users of changes made, or contemplated.

Of particular interest to users of files is the ordering of the records. This, of course, is described in the schema, but special consideration should be given to this feature of a file which may enhance its usefulness greatly for a variety of purposes and is related to making the optimum and most economical applications on the particular computers employed.

Let us define some basic concepts and mention some elementary facts about some of them:

1. <u>Key, sort key, or retrieval key</u>. The identifying field.

2. <u>Collating sequence</u>. The particular order that sequencing follows.

3. <u>Sequential order</u>. Arrangement of a file so that the key field is arranged according to the collating sequence. If a file

organization is sequential, records are stored and accessed consecutively. Advantage - rapid access to next record. Disadvantage - difficulty in correcting or updating the file.

4. <u>Relative random order</u>. According to a particular attribute, the file is in random order, though it is ordered according to the sort key. (For example, the visual magnitudes of stars in the Bright Star Catalog which are arranged by BS=HR numbers are in relative random order.)

5. <u>Random order</u>. The location of a record is random on the access device, though obtainable mainly through (a) dictionary lookup, or (b) calculation (key is mapped onto an address). Advantage - a record may be retrieved in a single access without disturbing other records, and thus, updating or correction of records is easy. Disadvantage - records are normally of equal length. Not rapid for accessing large numbers of records. Overflow problems arise (different records are mapped to same address). Large dictionaries may be necessary.

6. <u>List structure (associative memory)</u>. Each record points to (contains the address of) the next record. There are many useful variations of list structures, such as, the <u>ring structure</u>, where the last record points to the first, the <u>coral structure</u>, where there are backwards as well as forward pointers, and the <u>hierarchical structure</u>, where particular attributes have their own pointers through the list or sublists. Advantage - extremely flexible arrangement of memory allowing for changes in list size and updating. Disadvantage - large overhead in managing such a system.

7. <u>Inverted file</u>. The file is ordered according to every attribute of interest which contains pointers to keys (data items become keys). Advantage - extremely useful to extract maximum information from files, particularly when inquiries of the file are unpredictable. Disadvantage - the dictionary may be larger than the data base itself and is difficult to manage and maintain.

Figures 2, 3, and 4 illustrate list and inverted structures using selected information from the Bright Star Catalogue.

PROBLEMS OF COMPATIBILITY

Problems of compatibility exist on many levels, the principal ones being:

1. languages

BS	Name		Double Star Catalogue	RA (2000)	DEC (2000)	Vis. Mag.	Sp. Class	Par
7476	54	SGR	12767A	$19^h40^m44^s$	-16°17'	5.45H	K2III	+".030
7477				19 37 57	+49 17	6.35R	dG6	+.006
7478	12 φ	CYG		19 39 23	+30 09	4.64R	G8III–IV	+.007
7479	5 α	SGE	12766	19 40 06	+18 01	4.30R	GOIII	-.004
7480	45	AQL	12775	19 40 43	- 0 37	5.52H	A2	+.014

Figure 2. Exerpts from Catalogue of Bright Stars.

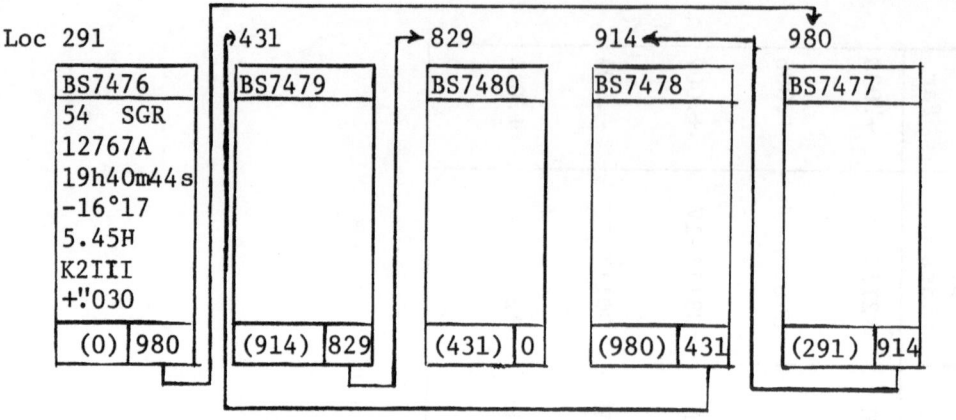

Figure 3. A list structure arrangement of the records in Figure
2. 0 indicates the end of the list. If the 0 were replaced by
291, this would be a ring structure. Backwards pointers are
shown in parentheses.

Name	BS
AQL 45	7480
CYG φ 12	7478
SGE α 5	7479
SGR 54	7476

Vis.Mag.	BS
V < 5	7478,7479
5 ≤ V ≤ 6	7476,7480
6 < V	7477

Double Star Cat.	BS
12766	7479
12767A	7476
12775	7480

Spectral Class	BS
A	7480
G	7477,7478,7479
K	7476

R.A. (2000)	BS
19h35m to 19h39m59s	7477,7478
19h40m to 19h44m59s	7476,7479,7480

Parallax	BS
π ≤ 0	7479
π < +.010	7477,7478
π ≥ .010	7476,7480

Dec. (2000)	BS
> 0	7477,7478,7479
≤ 0	7476,7480

Figure 4. Inverted file structure for the file of Figure 2.
Data Items in the Bright Star Catalogue serve as keys in the
inverted file.

2. machines or devices.

Language compatibility is, for the most part, a superable problem
since the machine-independent development of compiler languages
like FORTRAN, ALGOL, COBOL and PL/I. Machine incompatibility,
however, has remained a serious problem, often necessitating
that data, machine-readable for a particular set of machines, be
re-written on media machine-readable on another set of machines.
Fortunately, computer manufacturers have moved more and more in
the direction of standardization, recognizing the enormous costs
involved in re-writing programs and data.

Still, an obvious modus operandi is available in many instances. Clearly, data files stored on data cells, disk packs,
drums and similar devices are not generally transferable to
other systems. However, most magnetic tapes in use today are
7-track or 9-track, compatible with and fitting the tape drives
of most machines. This means, for several reasons, that tapes
play one of the dominant roles in data storage and transfer.
Data bases on data cells, disk packs, etc. usually have tape
media as backup in the event of corruption of the base. Even if
they do not, information on the data cells, disk packs, etc. can
be written on tape and thus become available to other users.

It used to be that the principal medium for storage was
punched cards, and, indeed, many of the astronomical data files
stored today in data centers are card decks. This basic medium
makes it possible to convert information to media which are
otherwise incompatible. In the worst case, where magnetic tape
compatibility does not exist, it will still be generally possible
to go from tape to punched cards and thence to compatible media.
It should be mentioned parenthetically that, in the original
compilation of files for a data base, it has been shown to be
more economical, as well as producing the least number of errors,
to use keytape rather than keypunch devices.

ASTRONOMICAL DATA BASE MANAGEMENT SYSTEMS

In building central astronomical data bases for concurrent
use of information through time-sharing or multiprogramming techniques, it will be necessary to have all files resident in the
same data base. With the methods described above, it should be
possible to have all the files available, even if there are major
differences in their structure. It then behooves the management
programs to access and manipulate the files so that inquiries and
requests can be made across all files. In the language of the
CODASYL Report, a "Data Description Language" containing the
schema, subschemas, and lock information must be developed for
the administrator; a "Data Manipulation Language", which is

procedural, must be developed for access to the Data Base; and, finally, a "Data Management Routine" must be developed to maintain and preserve the integrity of the Data Base.

In operating an Astronomical Data Base Management System, a new class of problems will be encountered, such as "deadlock", when two or more run units are queued and each competes for the unit held by the other, but the considerable experience of others in the field of Data Base Management Systems will make the transition to such a system relatively smooth.

REFERENCES

Bassler, R. A. and Logan, J. J.: 1973, The Technology of Data Base Management Systems, College Readings, Inc., P. O. Box 2323, Arlington, Va., 22202.

Bureau International D'Information sur les Ephémerides Astronomiques, Fiches d'Information N°1 à 125, Palais de L'Institut, 3, rue Mazarine, Paris VI°, France.

CODASYL: 1969, Data Base Task Group Report, October 1969 (out of print).

CODASYL: 1971, Data Task Group Report, April 1971, ACM, 1133 Avenue of the Americas, New York, N. Y., 10036.

CODATA: 1868-1976, CODATA Newsletter 1-16 (twice per year), CODATA Secretariat, 51 Bd. de Montmorency 75016, Paris.

CODATA: 1969, International Compendium of Numerical Data Projects, Springer Verlag, Berlin, N. Y.

CODATA: 1969-1973, CODATA Bulletin 1-11 (irregular), CODATA Secretariat, 51 Bd. de Montmorency, 75016, Paris.

Comrie, L. J.: 1928, On the Construction of Tables by Interpolation, Monthly Notices, R.A.S., Apr. 1928.

Eckert, W. J.: 1940, Punched Card Methods in Scientific Computation, The Thomas J. Watson Astronomical Computing Bureau, Columbia University, N. Y.

Flores, I.: 1970, Data Structure and Management, Prentice-Hall, inc., Englewood Cliffs, N. J.

GUIDE-SHARE Data Base Requirements Group: 1970, Data Base Management Systems Requirements, SHARE Secretary, Suite 750, 25 Broadway, New York, N. Y., 10004.

Hondius, F. W.: 1975, Emerging Data Protection in Europe, North Holland Publ. Co., Amsterdam.

House, W. C.: 1974, Data Base Management, Petrocelli Books, New York.

International Computer State of the Art Report 15: 1973, Data Base Management, Infotech Information Limited, Maidenhead, Berkshire, England.

IAU, Proceedings of the 14th General Assembly: 1971, Proposal for the Establishment of an International Information Bureau on

Astronomical Ephemerides, p. 84, D. Reidel Publ. Co., Dordrecht, Holland.

Katzan, Jr., H.: 1975, <u>Computer Data Management and Data Base Technology</u>, Van Nostrand Reinhold Co., N. Y.

Klimbe, J. W. and Koffeman: 1974, <u>Data Base Management</u>, North Holland Publ. Co., Amsterdam.

London, K. R.: 1974, <u>Documentation Standards - Revised Edition</u>, Petrocelli Books, N. Y., pp. 171-177.

PRINCIPLES OF A CODED NUMBERING SYSTEM AND ITS APPLICATION TO OPEN CLUSTERS

J.C. Mermilliod

Institut d'Astronomie de l'Université de Lausanne
and Observatoire de Genève, Switzerland

When we began to collect systematically photometric data it was felt that we needed an appropriate system of identification for our catalogues on magnetic tape.   The current system of identification which uses catalogue or list name and star number is best suitable for printed catalogues but brings out many drawbacks when using and intersecting catalogues on magnetic tapes.   Thus we developed a coded system of identification with the two basic conditions that it allows to include in the files stars with all possible kinds of numbering found in the literature, and that computer use was as simple as possible (Mermilliod, 1973).

One criterion of quality of a catalogue is that the stars are recorded under a single identification.   Because of the exotic and complex heterogeneity in the identification of stars found in the literature, star cataloguing brings forward many problems. Due to the anarchic development of identification systems we are obliged to define priority orders among the catalogues and to alter the original designation found in the literature to give our compilation a high degree of homogeneity in the identifications. For example, photometric data for a proper motion star can be found in the literature under the following identifications: HD, DM, Gliese, Luyten LTT, Giclas G stars.   Thus much secondary material is necessary in the form of transit tables and cross-identifications.   This problem has to be solved separately for field stars and for stars in open clusters.

This system was first used for the photometric files prepared in the Lausanne and Geneva observatories for the CDS.

C. Jaschek and G. A. Wilkins (eds.), Compilation, Critical Evaluation, and Distribution of Stellar Data. 19-22.
Copyright © 1977 by D. Reidel Publishing Company, Dordrecht-Holland. All Rights Reserved.

SCHEMATIC PRINCIPLES AND MEANING OF THE CODED NUMBERS

A coded number presents 10 digits.   The first is a sign
which separates southern DM stars from the northern ones.   For
the rest of the codes, it is always positive.   The general meaning of a code is contained in the second digit which runs from 0
to 9.   This is the first key to decode a coded number into the
two parts contained in the 8 remaining digits: a) the second key
which gives the reference to a catalogue or a published list, b)
the number of the star in the quoted reference.   The schematic
definitions of the coded numbering system are given in the appendix.

SOLUTION FOR THE FIELD STARS

Among the 73,000 measurements compiled in the UBV system
(Mermilliod and Nicolet, 1976), some 46,000 concern HD, HDE and
DM stars, 16,000 stars in open clusters, and 12,000 remain stars
with various identifications.   Some of them come from well-known
catalogues, such as the Gliese catalogue (Gliese, 1969) or the
luminous stars in the southern milky way (Stephenson and Sanduleak,
1971), for example.   Such catalogues are very valuable and all
stars contained in them having no other identification than that
of a specific catalogue, are easily coded.   So also are stars
from the selected areas.   But there are in the literature plenty
of lists with less than 100 stars, or less than 1,000 for more extended survey.   One good example is the Feige stars (114 in total)
(Feige, 1958) and another could be a sequence of stars in a chosen
constellation.   Such lists have their utility too and should not
be neglected during the compilation of large files.   In the framework of the coded numbering system a solution to each problem has
been looked for.

One major advantage of this system is that no accurate coordinates are necessary to form the files.   With the references,
one can easily go back to finding charts.   A second one is that
stars of a catalogue are recorded together in a file and not dis-
persed throughout it.

STARS IN OPEN CLUSTERS

Up to now, stars in open clusters were the most difficult to
include in large compilations, principally because there exist for
each cluster many numbering systems.   This results in an intricate
situation, whose solution needs an appreciable amount of work and
material.   The general form of a coded number for a star in an
NGC cluster is +2, followed by the NGC number and the star number
in the cluster.   This is simple, but the star's number is difficult to define, due to the multiplicity of the available numbering

systems.   In order to solve correctly this problem we need cross-
identification tables between different numbering systems and also
between the adopted final system and the HD and DM catalogues,
because bright stars in open clusters are frequently included in
lists of field stars under their HD number.   These realizations
are explained in more detail and with references in the Catalogue
of UBV photometry and MK spectral types in open clusters (Mermilliod, 1976).   As for field stars, we note that stars of an
open cluster are recorded all together in a specific place in a
large file, and thus their usage is greatly facilitated.

CONCLUSION

Taking into account all the definitions and principles, one
obtains eventually quite a homogeneous numbering system which
helps to keep up to date high quality catalogues which we hope
contain only a small number of redundant identifications.

Computer use has proved to be very effective and the more
catalogues are coded with this system, the more easy is the comparison of many catalogues, photometric, spectroscopic and so on.

REFERENCES

Feige, J.:   1958, *Astrophys. J.* 128, 267.
Gliese, W.:   1969, *Catalogue of Nearby Stars*.
Mermilliod, J.C.:   1973, *CDS Internal Report* No. 5.
Mermilliod, J.C.:   1976, *Astron. Astrophys. Suppl.* 24, 159.
Mermilliod, J.C. and Nicolet, B.:   1976, in preparation.
Stephenson, C.B. and Sanduleak, N.:   1971, *Publ. Warner and
    Swasey Obs.* No. 1.

APPENDIX

Schematic definitions of the coded numbering system

Durchmusterung catalogues

±0.0ZZBBBBB   ZZ: declination zone   BBBBB star number

HD and HDE catalogues

+1.00BBBBBB   BBBBBB star number

Open clusters

+2.NNNNBBBB	NNNN NGC number	BBBB star number
+3.IIIIBBBB	IIII IC number	BBBB as above
+4.AAXXBBBB ⎫	Anonymous clusters	⎰ AA,A cluster names
+5.AXXXBBBB ⎭		⎱ XX,XX cluster no.
		BBBB star number

Star catalogues and star lists

```
+6.0AABBBBB ⎤
+6.1AAABBBB ⎬   AA, AAA, AAAA author identification
+6.2AAABBB  ⎦   BBBBB, BBBB, BBB star number
```

Selected areas

```
+7.AXXXBBBB    A   catalogue of selected areas
               XXX S.A. number
               BBBB star number
```

Catalogues by declination zones (except DM catalogue)

```
+8.AAZZBBBB    AA author identification, ZZ declination zone
               BBBB star number
```

Miscellaneous lists

```
+9.0AAABBBB    Stars in associations

+9.5RYXXBBB    R = 0 : Harvard standard regions
               R = 1 : Cousins E regions
               Y = 1-6, stands for A-F
               XXX = 01-09 : region designation
               BBB  star number
```

THOUGHTS ON THE FORM OF PRESENTATION OF STAR CATALOGUE DATA

H. Eichhorn

Universitätssternwarte Graz, Graz, Austria

ABSTRACT

The reductions of the direct measurements necessary for obtaining the published positions and proper motions of stars are rather involved. The transformation from equatorial coordinates to galactic coordinates – and proper motion components – is rather simple, and due to the exact definition of the galactic system, galactic coordinates and proper motion components can be given with the same accuracy as equatorial ones. Since it is data in the galactic system which are primarily of interest for the galactic astronomer, it is proposed that much data be henceforth published in catalogues of star positions and proper motions. One may assume that the minority of astronomers who need the data in the equatorial system will be least likely to make mistakes in transforming them back to the more direct form in which they were observed.

It is further suggested that the SAOC could be used as a base for continuous improvement and updating of a comprehensive catalogue of star positions, to be updated by the incorporation of newly available material every other year or so. Procedures by which this would be accomplished are indicated.

C. Jaschek and G. A. Wilkins (eds.), Compilation, Critical Evaluation, and Distribution of Stellar Data. 23.
Copyright © 1977 by D. Reidel Publishing Company, Dordrecht-Holland. All Rights Reserved.

A BIBLIOGRAPHY OF ASTRONOMICAL CATALOGUES AND THE ORGANIZATION OF STELLAR DESIGNATIONS

M.J.Collins

INSPEC, The Institution of Electrical Engineers,
Hitchin, England

ABSTRACT

A comprehensive bibliography of astronomical catalogues for the period 1951-75 inclusive containing over 2000 items and including an index of designations has shown that there is an urgent need to control the use of designation notation. It is suggested that a small data base, similar to the one created for this bibliography could serve as a useful guide to the assignment and subsequent citation of catalogued objects. This in turn would help those compiling inverted files of stellar objects such as the Catalog of Stellar Identifications.

INTRODUCTION

A computer file of astronomical catalogues covering the period 1951-75 inclusive has been constructed replacing a simple card file used by INSPEC information scientists to identify catalogue designations used in the literature. A bibliography, soon to be published from the computer data base, contains well over 2000 entries and includes a designation index. This index is believed to be the first of its kind to cover all areas of astronomical research and lists over 1200 items.

AMBIGUOUS DESIGNATIONS

That there can be so many designations in only a 25-year period suggests that we may be getting into some difficulty and indeed my designation index points to some very awkward

ambiguities that have arisen in the literature. Assuming, therefore, that the same designating letters have been used in two or more catalogues, certain things can be said. Some ambiguous cases although undesirable can to some extent be justified:
a) when two catalogues, similarly designated, refer to completely different objects, e.g. if there was a radiosource list using " P-L " for its designation then there would be confusion with the Palomar-Leiden survey of faint minor planets but the context of their use would obviously decide which P-L list was being used
b) when the same designation is being used but the 'format' is different, e.g. P-L 1234 would almost certainly come from a different list to P-L 12-34-56, and
c) when the authors are using a possibly ambiguous designation but state which catalogue they are using.

However, confusion arises:
i) when authors fail to state which catalogue they are using, and
ii) when (c) applies but somebody else takes their data out of context and uses it elsewhere!

It is this last case that I wish to consider here, particularly as it applies to the compilation of inverted files of stellar data using information appearing in the literature.

UN-DESIGNATED CATALOGUES

The catalogue without a designation is perhaps the biggest single cause of ambiguity. Subsequent papers may cite an un-designated catalogue in all sorts of ways, usually the original author's name or initials become involved, but it is not for several years that some sort of recognizable designation emerges. This period is when the confusion develops. There are additional traps. Sometimes the author does propose a designation but later realizes that it is already in use. The LkHα notation, formerly the ambiguous designation LHα, is a good example where things turned out alright in the end!

Let us look at one particular example.

In Fig. 1 the star designated W175 appears in the photometric and polarimetric study by Carrasco, Strom and Strom (1975). The star is in fact number 175 in the catalogue of stars in the young cluster M16 (NGC 6611) by Walker (1961). Now Walker did not specify how his objects should be designated so W175 is quite a reasonable label especially as Carrasco and his colleagues clearly state where their W-designated stars come from.

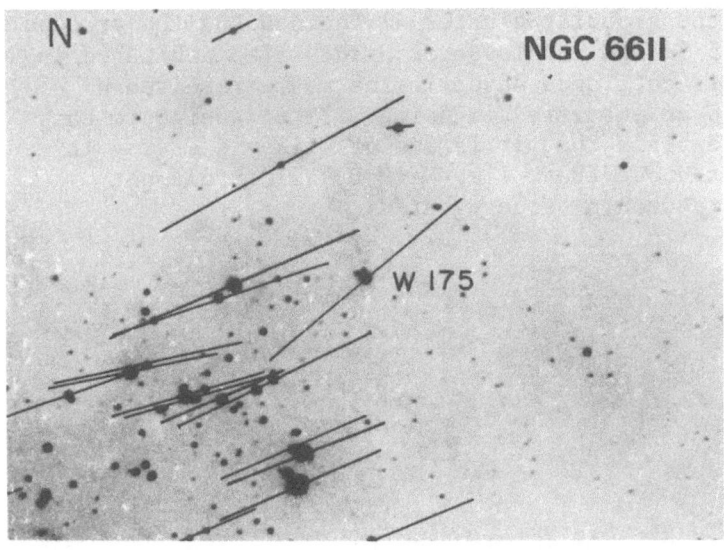

Fig. 1: W175 from Walker's catalogue of stars in NGC 6611.

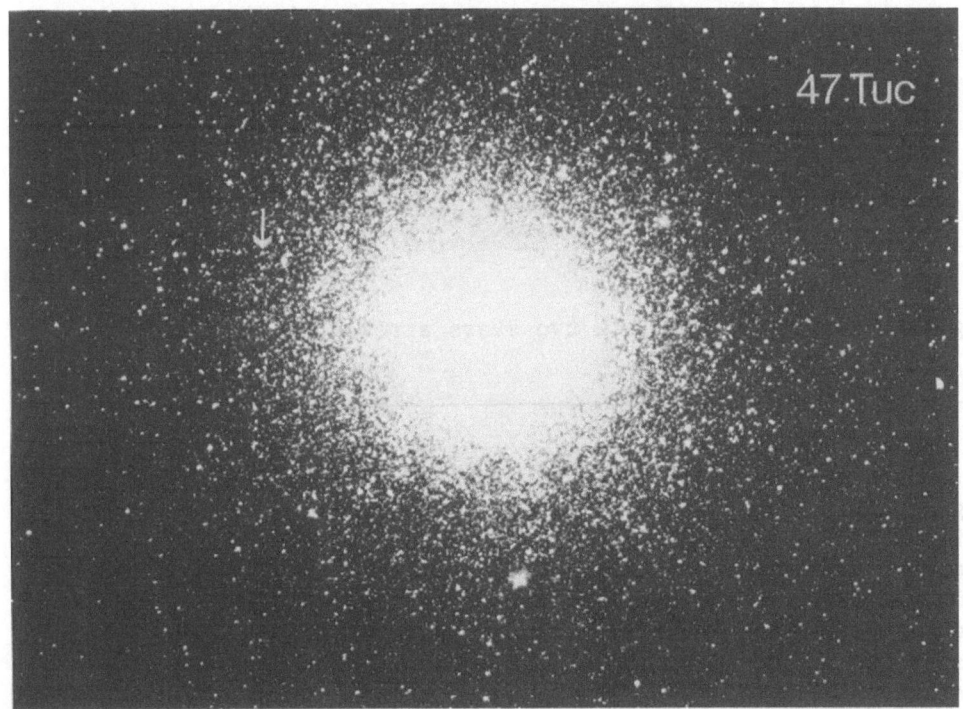

Fig. 2: The star arrowed at left of centre is W175 from Wildey's catalogue of stars in 47 Tucanae.

If we now start looking for ambiguities we have not far to
look. Fig. 2 shows the globular cluster 47 Tucanae and the arrowed
star is numbered 175 in the catalogue of Wildey also published in
1961. Stars from this catalogue have, in the past, received W
numbers and we could be justified in using W175 to designate this
object provided we explain that it is one of Wildey's stars. It
should also be pointed out that, like Walker, Wildey did not
specify how his stars should be designated.

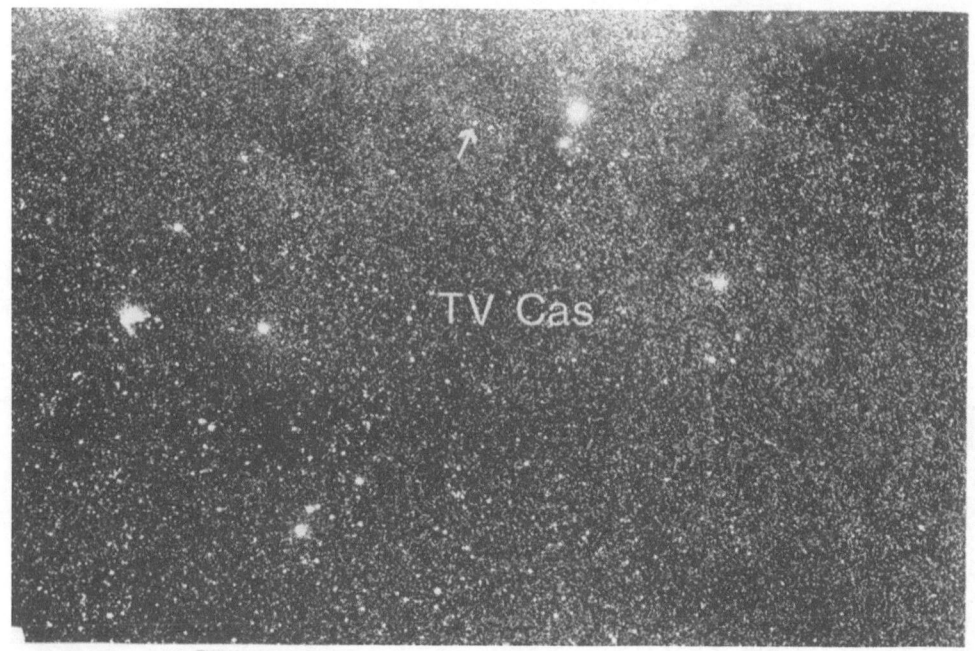

Fig. 3: The brighter of the two stars arrowed is the variable
TV Cassiopeiae.

Finally let us consider TV Cassiopeiae (HD 1486) shown in
Fig. 3. This star was found to be variable way back in September
1911 by Astbury but 42 years later it was listed in Wilson's
'General Catalogue of Stellar Radial Velocities' (1953) - as
number 175 !! Now stars from the GCSRV have in the past also
been given W numbers so we end up with three possible stars all
designated or at least all with a possible chance of being
designated W175.

DESIGNATION CONTROL

The problem of keeping three such designations separate on a data base which is collecting information relating to individual stars is of course obvious. Work done already on the Catalog of Stellar Identifications and the Bibliographical Star Index has shown that these ambiguities can be alleviated by using, for example, alternative designations such as HD numbers. I suggest that more control in the assignment and citation of star designations may prevent the need for such 'mopping up' operations by the data base producers. It must also be remembered that ambiguity case (a) may start to cause problems following the rapid increase in the number of stars identified with objects originally designated as radio or X-ray sources. I further suggest that a small data base, dealing only with designation use, similar to the one created for my bibliography would be enough to control the situation if it were used. Authors would be able to look at a list of designations already in use and Walker, had he had such a list back in 1961, could have suggested 'WALKER' or perhaps 'WALK' for his stars instead of running the risk of them becoming just W objects.

REFERENCES

Carrasco,L., Strom,K.M., Strom,S.E.: 1975, Rev. Mex. Astron. & Astrofis. 1, 283
Walker,M.F.: 1961, Astrophys. J. 133, 438
Wildey,R.L.: 1961, Astrophys. J. 133, 430
Wilson,R.E.: 1953,'General Catalogue of Stellar Radial Velocities' Carnegie Inst. Washington Publ. no. 601 = Pap. Mt. Wilson Obs. 8

REFERENCES FOR FIGURES

Fig. 1: Carrasco, Strom and Strom, 1975, Rev. Mex. Astron. & Astrofis. 1, 283, Plate 11
Fig. 2: Photograph from the Mount Stromlo Observatory (available from the Royal Astronomical Society, Ref. no. RAS 621)
Fig. 3: Photograph from the Lowell Observatory (available from the R.A.S., Ref. no. RAS 496)

THE CATALOGUE OF STELLAR IDENTIFICATIONS

F. Ochsenbein, D. Egret, M. Bischoff

Centre de Données Stellaires, Observatoire, Strasbourg (France)

1. DESCRIPTION OF THE CATALOGUE

Collecting all the data and references for a given star is generally a very time-consuming task, because the star may have many identifications. It is therefore a very first need to know the various identifications which may be used in the primary literature concerning a given star, in order to retrieve the whole set of information about the star.

The Catalogue of Stellar Identifications (CSI), which tries to solve this problem of cross-identifications, is therefore the fundamental file of the Stellar Data Center. It was created five years ago by J. Jung (1971), who was the first Director of the Center, with the help of Madame Bischoff and Madame Morin. The CSI was first a merging of some fundamental catalogues: SAO, HD, AGK2, CPC (Cape Photographic Catalogue for 1950), YZ (Yale Zone Catalogue), and GC. For each star, the information was the following:
- 1950 coordinates, rounded to $2^{-20}*360°$ (roughly $1''3$)
- magnitudes, visual and photographic (precision: 0^m1)
- spectral type taken from HD
- DM identification (BD, CoD or CPD)
- identifications in HD, AGK2, CPC, YZ, GC.

During the course of this task, errors in the fundamental catalogues were detected and corrected, and correspondences between DM identifications in the zones −22° and −40° to −52° were established.

C. Jaschek and G. A. Wilkins (eds.), Compilation, Critical Evaluation, and Distribution of Stellar Data. 31-36.

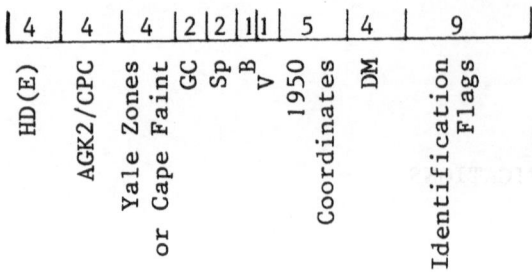

FIGURE 1.   Description of a logical record of the CSI.

   New catalogues were then connected to the CSI. Many difficulties arose in the course of this merging task, always related to the stellar identifications, especially with the Index
Catalogue of Visual Double Stars. To avoid the difficulties
encountered with a very large data base of varying-length records,
only a 1-bit flag indicates that a given star belongs to one given
catalogue; a secondary file provides the identification of the
star in the catalogue concerned.

   Fig. 1 describes the structure of the main index, the CSI.
For each star, the information is stored in 36 bytes records.
HDE identification is provided when the star is listed in vol.
100 or plotted in vol. 112 of Harvard Annals; this identification
work was achieved by R. Bonnet (1975). The magnitudes were homog-
enized to (UBV) (Ochsenbein, 1974). The coordinates are not
homogeneous; the majority of them are taken from SAO, but for
about 40 000 stars, the precision is about 1' (HD coordinates).
The DM identification, which represents actually the keyword of
the star, has been extended to non-DM stars. It can be seen that
up to 72 catalogues can be flagged in the CSI; at present, the
number of connected catalogues is now roughly 30; Table 1 lists
these catalogues.

   The CSI itself now contains roughly 440 000 stars and is
thought to be complete up to 9^m5 visual.

   For the present time, the data base is developed on the INAG
computer located at the Meudon Observatory, to which the Stellar
Data Center has a remote access.

2.   FACILITIES PROVIDED BY THE CENTER

   The general structure of the data base is described in Fig. 2.
The fundamental catalogue, the CSI, is connected to the secondary
files; all these files are on direct-access devices. These secondary files are:

TABLE 1. Catalogues connected to the CSI.

Catalogue	Data	Number of Objects
SAO AGK2/3 Yale Zones GC	Proper motion	360 000
FK4	Proper motion	2 000
N30	Proper motion	5 300
GC Trig Par.	Parallaxes	5 800
Blanco et al. Mermilliod	UBV UBV	26 000
Geneva	7 color phot	5 000
Hauck et al.	uvby β	9 600
Infrared Sky Survey	2 u phot	5 600
Celescope	UV	5 700
Variable stars		3 000
HD	Sp. types	260 000
Jaschek et al. Kennedy	MK Sp. t.	23 000
Wilson Abt and Biggs	Rad vel	25 000
Uesugi and Fukuda Bernacca	V sini	4 000
Luminous stars		9 000
HR		9 100
ADS/IDS		40 000
Batten	Sp. bin. orb.	730

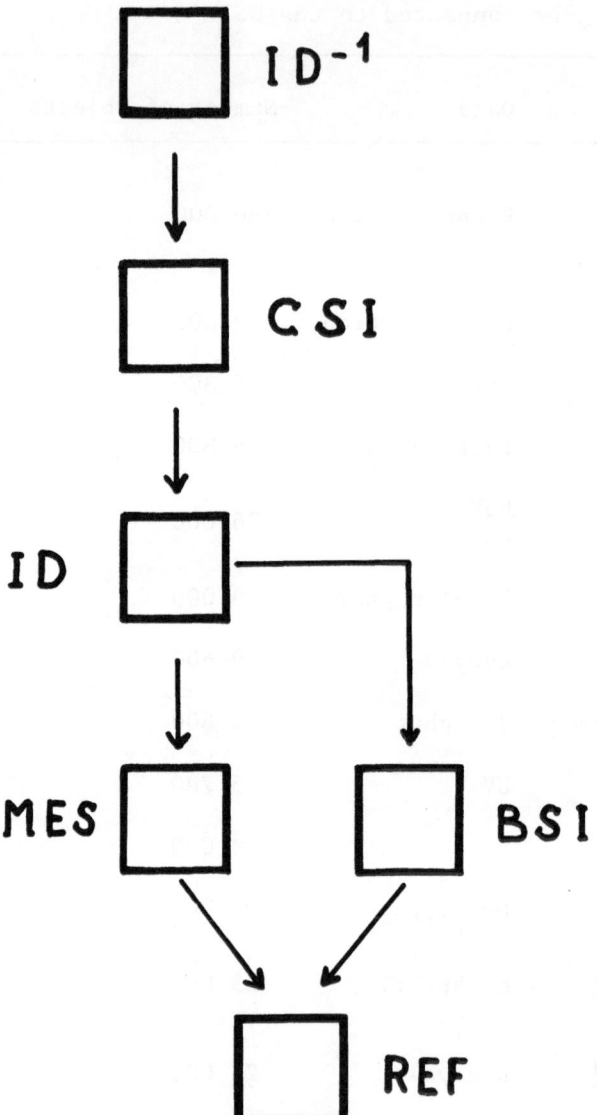

FIGURE 2. Schematic description of the data base.

- ID: provides the identifications of the stars which are flagged
   in the CSI

- ID^{-1}: provides the DM identification from the identification
   taken from one of the connected catalogues - the reverse
   file of ID

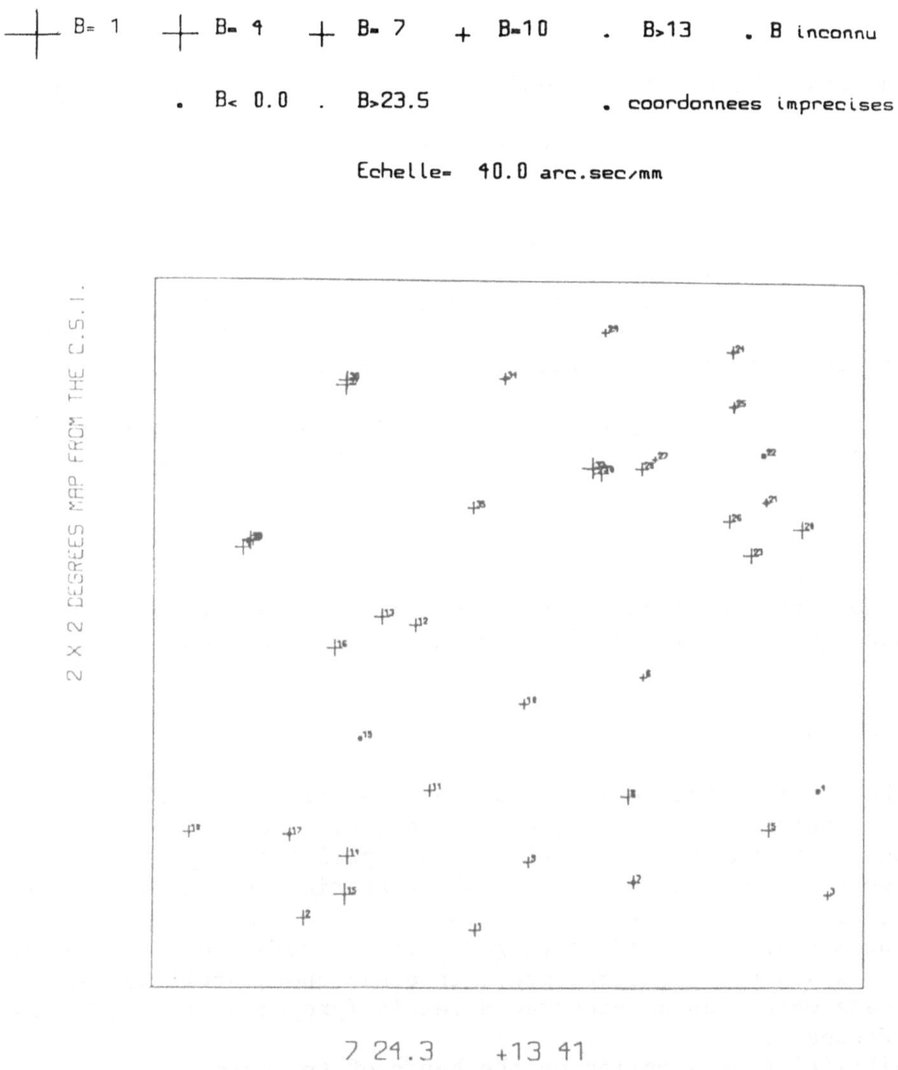

FIGURE 3. Example of a plot of CSI stars (field 2 x 2 degrees, centered at $\alpha_{1950}=7^h 24^m3$, $\delta_{1950}= +13°41'$.

- MES: provides some of the data which are known about the stars
    - in (UBV), (uvby), Geneva photometries
    - in MK classification
    - in infra-red photometry (2 microns sky survey)
    - in trigonometric parallaxes
    - in radial velocities

- BSI - Bibliographical Star Index - which will be described by F. Spite later on and provides the bibliographic references for each star

- REF: bibliographical references, listing the titles of the
  papers

On this scheme, the following facilities have been developed:

a - Retrieval of all the information for a given set of stars
    defined by any identifiers

b - Creation of a subset of the CSI on the basis of given criteria,
    such as position in the sky, magnitudes, spectral type, or
    combinations of data requirements. This facility is especial-
    ly suited for statistical analysis, which requires the whole
    information known at present for a combination of data.

c - Plot of the stars lying in a given field of the sky; this is
    especially suited to the identification of stars on a chart
    or the preparation of observations. Fig. 3 illustrates this
    facility.

For each of these facilities, all informations (identifications, coordinates, magnitudes, spectral type, data taken from
the catalogues, bibliographic references) are provided.

3. FUTURE DEVELOPMENTS

The possible future extensions of the CSI will probably be:
- the extension of the connected catalogues, since many of the
catalogues are updated regularly by the specialists (photometric
catalogues at the Lausanne-Geneva Observatories, radial velocities
at Marseille); we try also to include the whole HDE;
- the number of connected catalogues, for example Michigan spectral
survey, catalogues of measurements in other photometric systems;
- the data which can be accessed directly (proper motions selected
by A. Fresneau);
- possibilities of sampling on the basis of the data.

We hope that the CSI will also become useful to people outside
our Data Center, because about five years of effort have been put
into it. We would be very thankful for all suggestions for the
improvement of the CSI.

REFERENCES

Bonnet R. and Ochsenbein F. (1975) Inf. Bulletin Strasbourg
      Center 8, 8.
Jung J. (1971) Inf. Bulletin Strasbourg Center 1, 3.
Ochsenbein F. (1974) Astron. Astroph. Supp. 15, 215.

DESIDERATA FOR THE CATALOGUE OF NEARBY STARS

W. Gliese

Astronomisches Rechen-Institut Heidelberg

In 1969 the second edition of the "Catalogue of Nearby Stars" was published. It contains 1529 single stars and systems with a total of 1890 components. The catalogue lists all stars with parallaxes equal to or larger than 0".045. Only 1049 of these objects are nearer than 20 pc. The "Catalogue of Stars within 25 pc of the Sun", published 1970 in the Royal Observatory Annals No. 5 consists of 1744 systems of which 1566 trigonometric parallaxes.

Since that time new programmes for determining trigonometric parallaxes have been started, concentrating mainly on nearby faint objects as white dwarfs and red dwarfs. The catalogues of the US Naval Observatory already have listed several hundreds of new distances determined with the 61-inch reflector at Flagstaff. Furthermore, some lists with spectroscopic and photometric parallaxes have been published. Altogether, about two hundred new stars at least can be added to the collection of objects in the solar neighbourhood. Therefore, the compilation of a third catalogue of nearby stars seems to be recommendable around 1980.

Such an undertaking consists of two parts:1) Collection of new data, and 2) selection of the best value of a quantity of which various measurements are known. In practice, very often the second task will be more problematic and more difficult than the first.

Let me restrict myself to a few basic questions:The well-known problems of a uniform system of trigonometric parallaxes, the weighting and combining of various measurements will be investigated by Dr. van Altena at Yale Observatory. I shall appreciate being able to use his results for the new catalogue of nearby stars.

C. Jaschek and G. A. Wilkins (eds.), Compilation, Critical Evaluation, and Distribution of Stellar Data. 37-39.
Copyright © 1977 by D. Reidel Publishing Company, Dordrecht-Holland. All Rights Reserved.

For spectroscopic and photometric distance determinations, especially of objects of low luminosity, the situation is more vague. Various "spectral type-luminosity" relations and "colour-luminosity" relations have been used by different authors. As compiler of a catalogue based on distances of stars, I would like to have a collection of these relations with a critical summary which would allow me to determine luminosities without wasting of time with additional investigations.

Basic sources of spectral types and photoelectric colours are MK classifications and UBVRI measurements. If different values are known for the same quantity of a star, we should not use the mean of all determinations. Normally, the catalogue compiler who has to restrict himself to one value only, will assign priorities to certain series of observations. But, usually, he does not consider himself an expert on such observations. As he can draw nearly no support from literature, he needs private advice from colleagues. The work of the Data Centers is of invaluably high benefit for the compilation of a catalogue of nearby stars. The addition of a system of priorities seems to be very useful but, I realize, that such a solution of the problem will be nearly impossible.

It is necessary that the stars listed in a catalogue can be identified unambiguously. For the large number of faint objects, the positions should be given at least to the second in RA and to 0!1 in Decl. If identification charts are published in the literature, references will be made.

Some of these faint objects are identified in the Astrographic Catalogues. A central collection of the most reliable values for the constants of AC plates would be appreciated.

No uniform system of the proper motions of faint objects can be given. But the situation is not very serious, since an error of 0!1 per annum corresponds to an error in the tangential velocity of only 1 km sec^{-1} at the distance limit of the catalogue members.

For radial velocities I hope to make use of the work done by Prof. D. Evans and by the CDS.

Summarizing, let me emphasize that the compilation of a third edition of the Catalogue of Nearby Stars will be decisively supported by the collections of data centers. If various determinations of the same quantity are known, I would appreciate preference being given to one value which may be the mean value or the most reliable measurement.

I have need of a center for spectroscopic and for photometric parallaxes or, at least, of a center for the various "spectral type-luminosity" relations and "colour-luminosity" relations.

For identification of very faint objects, observers should give accurate positions or identification charts with their epochs.

Last, not least, there should be a priority list for the different systems of nomenclature.

The English version of the text was graciously corrected by Dr. R. Scholl.

A PROPOSAL FOR NON-AMBIGUOUS DESIGNATION OF STARS AND STELLAR
OBJECTS

F. Spite
Observatoire de Paris-Meudon

Summary : The number of new lists of stars and stellar objects
published each year increases rapidly. From the designation of a
star it is often difficult to trace back the original list or the
basic data relevant to the star or stellar object. Designations
are sometimes ambiguous. A few proposals are presented to try to
improve the present situation.

The number of new lists of stellar objects or stars appearing
in the literature increases from year to year with an amazing rate
and these lists are not always properly refered. It is therefore
more and more difficult for the astronomer to establish cross-references, especially when he is gathering information in a field
which is not his own specialized field. For instance the nomenclature of X-Ray Sources is somewhat confusing (Dolan, 1976). On
the other hand it is more and more useful to gather information
from various specialized field. An effort has therefore to be made,
to keep as clear as possible the designation of stars and stellar
objects.
    To illustrate the problem, may I quote the designation "W175"
which in addition to the identifications quoted by Collins (see his
paper in this Colloquium) may represent the star Wolf 175 (Wolf,
1913) the star Woolley 175 (Woolley, 1970) or the star Wray 175
(Wray, 1966).

After discussion with F. Ochsenbein, who is frequently con-
fronted to this problem at the Centre de Données Stellaires (Stellar
Data Center) in Strasbourg, I conclude that the simplest rule (and
therefore the first step to make) is the following : when a star or
stellar object is quoted in a paper from a number in a list or a

C. Jaschek and G. A. Wilkins (eds.), Compilation, Critical Evaluation, and Distribution of Stellar Data. 41-42.
Copyright © 1977 *by D. Reidel Publishing Company, Dordrecht-Holland. All Rights Reserved.*

catalogue, the list or the catalogue should appear with its complete
bibliographic reference among the other references.

Of course, I don't advocate the appearance of the complete
reference of the Henry Draper Catalogue, as soon as an HD star is
quoted. But why? Because this catalogue is very well known. So that
the complement of this rule follows by itself : the basic lists or
catalogues are gathered in a table, with approved abreviations,
worked out in such a way as to avoid any ambiguity. This would
enable to quote stars and stellar objects in very well known
lists or catalogues without writing down the complete bibliographic
reference (and even somebody not knowing astronomy enough to know
the Henry Draper Catalogue, could have there a way for learning
what the mysterious initials HD are meaning!). In my opinion,
Commission 5 of I.A.U could build such a table of catalogues and
abreviations. Then any designation of a star or a stellar object
could easily be traced back : either from the I A U table or from
the bibliographic references at the end of the paper.

Of course, the I.A.U. table can be easily increased, each
3 years for instance, so that any catalogue or list becoming
"basic" can be added to the list of the basic catalogues.

This proposal is not the complete solution of the problem,
but only a first step. Of course, this step could be improved.
For example, rather than the original list from which the star is
named, another list giving more precise data or more data could
as well be recommended: for instance, a proper motion star of the
list of Wolf has probably better parameters in modern lists than
in Wolf's original list, although it retains its name after Wolf's
list. It is always possible to add a note refering to a catalogue
better than the original one, either in the I.A.U table or in the
bibliographic references. But this is another problem, and I think
that the main one is to avoid any ambiguity, and for keeping with
simple things I suggest to stay with the proposal which could be
summarized by : you are urged to write down the references of the
lists or catalogues which are quoted in your papers and are not
yet basic enough to appear in the I.A.U table. Some practical
problems remain, such as "how to build up the I.A.U table". This
is to be decided upon agreement between the astronomers, but for
the stellar catalogues, a first basis for the discussion could be
the table of the catalogues which appear as the most frequently
quoted in the statistics of the Strasbourg Stellar Data Center.

Dolan, J.F. 1976, The Observatory, 96,66.
Wolf, M. 1913, Ver. Sternw. Heidelberg, 7, n° 10.
Woolley, R.v.d.R., Epps, E.A., Penston, M.J., Pocock, S.B. 1970,
Roy. Obs. Ann. n° 5.
Wray, J.D. 1966, Dissertation, Northwestern University, Evanston,
Illinois, reproduced by University Microfilms, Ann Arbor, Michigan.

THE USE OF SI UNITS IN ASTRONOMY

G. A. Wilkins

Royal Greenwich Observatory

SUMMARY

The principal features of the International System of Units (SI) are briefly described and the advantages of its wider adoption in astronomy are discussed. The need for the continued use of a system of astronomical units for some purposes is recognised.

1. THE ADOPTION AND NATURE OF SI UNITS

The International System of Units (SI) was adopted in 1960 by the General Conference on Weights and Measures to provide a practical system of units of measurement suitable for adoption in all countries for science, technology and general purposes. Details of the system are available in several official publications (see, for example, references 1 and 2), in recommendations to authors by international unions, scientific societies, and journals (eg refs 3, 4, 5) and in many textbooks. SI units are taught in schools and universities in many countries, and the system is now in widespread use.

There are three classes of SI units:- base units, derived units and supplementary units. There are seven base units: metre (m), kilogram (kg), second (s), ampere (A), kelvin (K), candela (cd), and mole (mol). The derived units are defined in terms of the base units, but may have special names and symbols (eg the unit of energy is the joule, J). The supplementary units are the radian and steradian, and these may for many purposes be treated as base units. In addition it is recognised that other units will continue to be used for some purposes; some of these are in general use (eg

the day and the angular degree), while others are in use in specialised fields (eg electronvolt, parsec). Other units have been
accepted temporarily (eg nautical mile, ångström, gal), but it is
hoped that SI units will be increasingly used instead. It is considered that some units of the centimetre-gram-second system
(c.g.s.) should no longer be used even though they may be directly
related to SI units and have special names (eg erg, gauss).

The system includes a set of names and prefixes which may be
used to form decimal multiples and sub-multiples of SI units. In
general these prefixes correspond to steps of 1000, but the factors
10 and 100 are recognised; compound prefixes are to be avoided.
For example, the centimetre (cm) and the cubic centimetre (cm^3)
may continue to be used; the name micron is replaced by the micro-
metre (μm), and millimicron by nanometre (nm).

2. UNITS FOR USE IN ASTRONOMY

In 1967 the International Astronomical Union adopted a resolution recommending "the general use by astronomers . . . of units
of the metric system . . .". It is understood that this was intended to be an endorsement of the use of SI units, but the resolution is ambiguous and so it has failed to give an impetus to the
change to SI units, rather than c.g.s. units. An examination of
current astronomical literature shows that many different sets of
non-compatible units are in use, sometimes even in one paper. The
wide variety of units is also noticeable at meetings; very few
listeners can give their full attention to the speaker while carrying out mental conversions of quantities from one system of units
to another.

It is almost certain that the coming generation of astronomers
will be familiar only with SI units and will wish to express all
their current results in these terms, even though they will have
to learn about the other systems in order to use the older literature. The use of SI units is already much more widespread in the
literature of physics and chemistry than it is in astronomy, and
their use is spreading in all fields of science, technology and
everyday life, even in countries where other units are still legal.
It is inconceivable that this trend will be reversed, and so it is
clear that it would be of general benefit if astronomers would
adopt the SI system, and use non-SI units only for certain recognised purposes, as quickly as possible.

There are three situations in which the use of non-SI units
is unavoidable in astronomy. Firstly, it would not be practicable
to use only radians and seconds for the measures of angles and
time, and so the sexagesimal units are recognised alternatives.
Secondly, the "natural" units for use in studying the dynamics of

the solar system are the mean distance of the Earth from the Sun,
the mass of the Sun, and the day. Even today, the mass of the Sun
is known in kilograms with less precision than are the masses of
the planets in units of the Sun's mass. Hence for this, and other
reasons, it is useful to define a system of astronomical units of
length, mass and time using a conventional value for the constant
of gravitation in these units. The relationships between these
units and the SI units are included in the IAU system of astronomical constants. New values of these constants were adopted at the
IAU General Assembly in 1976; the day is defined in terms of the
SI second. Thirdly, in stellar astronomy it is recognised that
the use of parsecs and magnitudes is both convenient and justifi-
able.

On the other hand, in physical studies of the planets, stars
and nebulae, the general adoption of SI units appears to be
extremely desirable. The use of the nanometre rather than of the
ångström and of the joule rather than the erg, to give two examples, involve only changes by powers of ten; it does not take long
to adjust to the fact that the density of water is 1000 kg/m^3.
The use of the electrical and magnetic units of the SI system does
require some changes in the formulae that are used, but the advantages of the adoption of a single system for all applications would
soon outweigh the disadvantages of the change.

3. CONCLUSION

I am hoping that it will be possible to include a firm recommendation for the general use of SI units in a "Guide to the presentation of astronomical data in the primary literature" to be
prepared under the auspices of IAU Commission 5. I would therefore
be grateful for information about any other non-SI units that are
necessary to meet the specialist requirements of astronomers. It
will be necessary to provide tables for the conversion between the
old and new units since much valuable astronomical data is contained in books, papers and catalogues in which the old systems
were used. The advantages of the use of SI units will become
readily apparent when the standard reference books and the principal astronomical journals all use a common system of units.

REFERENCES

1. Bureau Internationale des Poids et Mesures, 1970. Le Système
   International d'Unités. OFFILIB, 48 rue Gay-Lussac, F 75,
   Paris 5, France.
2a. National Physical Laboratory, 1970. SI The International
    System of Units. Her Majesty's Stationery Office, London,
    England.

2b. National Bureau of Standards, 1970. SI The International
    System of Units. NBS Special Publication 330, US Government
    Printing Office, Washington DC, USA.
    Note: 2a and 2b contain the same translation of reference 1.
3. CODATA, 1973. Guide for the presentation in the primary
    literature of numerical data derived from experiments. CODATA
    Bulletin No. 9. Available from CODATA Secretariat, 51 Boule-
    vard de Montmorency, 75016, Paris, France.
4. Royal Society, 1975. Quantities, Units, and Symbols. 2nd
    edition. ISBN 0 85403 071 9. Royal Society, 6 Carlton House
    Terrace, London, SW1Y 5AG, England.
5. Markowitz, W M, 1973. SI, The International System of Units.
    Geophysical Surveys 1, 217-241.

PART II

ACQUISITION AND PROCESSING TECHNIQUES

THE INFLUENCE OF ACQUISITION TECHNIQUES
ON THE COMPILATION OF ASTRONOMICAL DATA

Gart Westerhout

Astronomy Program
University of Maryland
College Park, Maryland 20742, USA

With the increasing use of electronic rather than photographic data collection, and on-line minicomputers at the telescope, two factors will necessitate a reconsideration of the way
in which data are stored at the telescope.
1) Many of the electronic data collecting methods produce data in a
final or semifinal (digital) form and usually with a large dynamic
range; high-data-rate digital recording devices at the telescope
are increasingly necessary.
2) Minicomputers allow quality control of data and often complete
data reduction either in real time or with a very short time delay.
We shall consider both these factors in some detail, emphasizing
some of the many possibilities inherent in the increasing availability of one- and two-dimensional array detectors and mass storage
devices. But let us first briefly examine the "classical" data
gathering techniques. As in the later discussion, we shall sub-
divide the field into high-resolution spectroscopy, low-resolution
spectroscopy (including multicolor photometry), surface photometry
of extended objects, and positional astronomy.

A. High-Resolution Spectroscopy. Usually high-resolution
spectra are obtained photographically at the Coudé focus of a telescope. The principal use of Coudé spectroscopy is for the study of
detailed line profiles in a variety of applications. Often, the
spectra are traced with a microdensitometer and usually digitized
in some form or another for work with stellar atmosphere models.
Neither spectra nor tracings end up in internationally accessible
data files, although people sometimes acquire older spectra from
observatory plate files. High-resolution spectroscopy is also used
in very detailed radial velocity work. It might eventually be useful to have in data files high-resolution spectra, in digital form,

C. Jaschek and G. A. Wilkins (eds.), Compilation, Critical Evaluation, and Distribution of Stellar Data. 49-59.
Copyright © 1977 by D. Reidel Publishing Company, Dordrecht-Holland. All Rights Reserved.

of all stars that have been investigated up to a certain resolution. Such a file, however, would be rather extensive considering the large amount of Coudé work that has been done already. Also many currently available spectra have never been calibrated to the extent necessary for real use in data files. With future digital recording techniques this situation might change, and in particular, it is to be hoped that Coudé investigators agree to some form of standardization.

B. Low-Resolution Spectroscopy. This is usually done at the Cassegrain focus of a telescope, sometimes at the primary focus, sometimes with objective prisms, and I would like to include filter photometry in this category. The main use of low-resolution spectroscopy is for stellar classification, emission line strengths, some radial velocity work, etc. Low-resolution data usually do end up in data files, in particular photometry data. Usually multicolor photometry information is available on tape for many stars and in many different systems and includes star numbers, rough positions and the several colors the author has measured. The usual transformation techniques are required in order to translate any set of data to a standard system, often a cumbersome program which requires a considerable amount of empirical determination of transformation constants. The difficulty with these kinds of transformation is just that: they are empirical and very often it is not quite clear why (physically) the various constants in the transformation equation have come out the way they are; moreover, the transformations are usually only valid for a relatively limited class of stars. Time and again the question arises whether such photometric data should be produced with reddening effects removed before it is stored on data files, or whether the reddening information should be contained in the data so that everyone using the data can try and correct in his own way. I am of the opinion that reddening corrections should be made before the data are finally stored in data files so that the true nature of the stars in question could be investigated without having to go through the routine time and again. Obviously, reddening data should accompany the color data of a star.

C. Surface Photometry. In this area we have two major subclasses. The first one is surface photometry directly at the telescope; there the data recording is similar to the recording of stellar data except that, instead of one data point per color, one gets an array of data points per color covering the extended object under consideration. I doubt that much, if any of such material is actually available in machine-readable form. The second category is photometry of photographic plates, which has increasingly been used in the study of extended objects, such as galaxies and emission nebulae. A photograph is scanned by a device which records all the intensities digitally, applies the necessary corrections to present as much as possible a linear intensity scale, and then

presents the data, often in the form of a contour map, sometimes
in the form of scans across the object in a number of different
directions. Although the investigators in this area have undoubtedly recorded their data on magnetic tape, they also usually have
published the results of their investigation in the form of contour
maps and the need for deposit of such data in an international data
bank does not seem to be very large at present. But more on this
below.

D. Positional Astronomy. In this area, there is a wealth of
data available in machine-readable form. The various positional
catalogs are all on magnetic tape, but these catalogs have been
arrived at after long and painstaking reductions of varied data and
are certainly not directly related to the data-taking technique.
If, in the future, positional astronomy gets to the point where a
telescope actually provides direct positions, the number of catalogs might well increase significantly.

It looks to me as though in all of these areas a very rapid
increase in data output will take place within the next 10 years.
I predict a flood of machine-readable data which in fact cannot be
used in any other way than through interaction with computers.

MINICOMPUTERS.

Let us consider the minicomputer at the observatory. This
versatile instrument enables the observer to monitor data-taking
processes, for example by assessing data quality, or deciding
whether enough data has been obtained for the required accuracy.
The minicomputer is able to perform a considerable number of on-
line data reduction tasks: determining raw magnitudes, (and since
it remembers data on extinction stars) monitoring of extinction and
updating of extinction corrections. There are many other tasks a
minicomputer can do, and it seems to me that the best way to illustrate this is by means of a few examples later in this paper.

There are already in existence fully automated multicolor
photometry systems in which the minicomputer follows a predetermined
observing program, setting the telescope, finding the stars, and
taking the data. If the computer also reduces all the data on-line,
the end product is a completely reduced set of colors and color
indices. Such a system would lend itself to a large scale survey
down to the limit of the telescope on which it is mounted. After
the initial investment in development time such a system will crank
out enormous quantities of data.

ARRAY DETECTORS.

But by far the largest problem I foresee will come when array
detectors come to be widely used. I will define an array detector
as a series of very closely spaced photon counting devices in an
array of anywhere between 100 x 100 and 1000 x 1000 of such
detectors. Such array detectors are currently under development
or being used experimentally in a number of places; some seem to
be no larger than a few tens of cubic centimeters. Somewhat less
extensive, often one-dimensional devices of this type are already
actively in use. As photon counting devices they have the same
sensitivity as our trusted photomultiplier and a high degree of
linearity over a very large range. It is clear that an array
detector will produce data at such a rate that entirely new storage
methods have to be used. This is a problem facing the radio
astronomers as much as the optical astronomers. The modern arrays
of radio astronomical antennas are able to produce maps of areas
of sky which measure typically 1000 x 1000 data points in a digital
form. In this case the data is usually obtained through computer
manipulation of data obtained by a number of telescopes in an
interferometer arrangement and the array of digital data is the
end result. The data is often displayed as contour maps although
radio astronomers increasingly have started to display some of
their data photographically in the form of a "radiophotograph".
The reason that such is often done in this area is because in some
of these pictures the dynamic range is not very large, that is,
intensities may differ only by a factor of 100 between highest and
lowest; at the same time a photograph often allows visual detection
of extended regions of extremely low surface brightness which dis-
appear in the noise when looked for in an array of numbers or a
contour map. But both in optical and radio astronomy we may well
have regions of sky where adjacent points have intensity differences of the order of 10^6. The array detectors we are talking
about and which are clearly in the offing are likely to have such
a large dynamic range. Therefore, recording of the output in the
form of a photograph, unless one does not choose to use this large
dynamic range, would be a complete waste of the capability of the
instrument; it would simply be used in the same manner as a normal
photographic camera with all the problems inherent in that. There
are two exceptions: one is logarithmic recording of intensities,
the other is the plates used in electronographic cameras, where
every photoelectron is recorded and in principle can be counted.

Let us consider the following scenario: Observe a galaxy or
emission nebula, using an array detector. The array detector
obviously produces, in the attached computer memory, a 1000 x 1000
point picture of the object studied. Since our imaginary device
is a photon counting device, it continues integrating, building
up signal. Suppose one integrates for one minute and puts the
output on a disk, then continues the integration, every minute

comparing the current integration with the first one. The purpose
of this is to detect interfering signals, noise, guiding errors,
etc. so that the computer, if necessary, either through action by
the operator or automatically, can discard all those one-minute
integrations which are not suitable. In the meantime, the picture
is building up on the screen of a CRT so that the observer can
watch whether he is getting the desired result. Obviously, the
observer is able to interact with the computer, so that for
example, he can occasionally display a scan across an interesting
region or enhance or decrease his contrast. Finally, we terminate
the exposure, put the integrated picture on magnetic tape and go
to the next object. Alternatively, a final series of reduction
steps could take place before this point: a) correct each point
in the array for gain (do not forget that each point is data
coming from the equivalent of one photo-detector) using a
pre-measured gain array, b) calibrate each intensity using a
pre-measured star or group of stars, so that the gain correction
is indeed on a proper, useable scale, and c) subtract background
by indicating on the CRT screen with some interactive device
which regions should be used for interpolation.

If a dynamic range of 10^6 is indicated and the array size is
1000 x 1000, this results in one million twenty-bit words, or a
total of twenty million bits (one-tenth of a reel of 1600 bpi tape)
for one picture! One might reduce this by restricting the recorded
data to the object only, which often will occupy less than half of
the picture (but many observers will want to retain all background
data!). One might further reduce this number, if appropriate (for
example, if the noise is statistical noise, i.e. goes up in absolute value with the strength of the signal), by reducing the word
length (which will usually be 24 bits rather than 20) to 16 bits,
still easy to handle by the computer. If one realizes that with
wide-band equipment, sky noise will limit exposures to one hour
at the very most, it is clear that extreme economy in data recording is needed.

Some do's and don't's are obvious: Always make sure that the
word length, even if it has to be variable from one problem to the
next, is appropriate for the problem at hand (i.e. make sure that
you only record significant digits), while at the same time it can
be handled easily by most computers. Try to avoid recording zero-
records if you can subtract them on-line. Do not be afraid of the
little bit of extra programming needed to read a format that is not
exactly standard. Almost every computer can read digital tape
regardless of its format; but translation programs from odd formats
to normal language might be expensive in computer time. Use
headers to define the record and to record information that does
not change with time, but do record all relevant information
(including comments on weather etc.) in the header. Documentation
of both observational information and of the reduction procedures

applied to the data is of utmost importance. For the scenario
described above, one might get away with half a reel of tape per
night. But it is also clear that with this type of equipment,
one might get all one's data reduction done at the time of the
observation so that the reduced data can be taken to a large computer for further analysis immediately after the observations.
Such analysis (not at the telescope) might consist of blowing up
certain parts of the picture, making scans across the picture in
a number of different directions, finding half-widths of features,
trying to represent parts of the picture by a combination of
gaussians, etc., etc. It may be that the observer is not very
interested, after he has obtained the results he was after, in
preserving the picture. Then the question is, is this picture
useful for other observers, and if so should the author go to the
trouble of keeping the many reels of tape he has probably amassed
during the operation of his fancy device? Such judgments are
extremely hard to make once one deals with numbers of data points
that are as large as I have envisaged in this example. But
eventually the maintenance of a data bank of digital images might
well become just as valuable as the use of deep 200-inch plates
by individuals other than the original observer.

MASS-STORAGE DEVICES.

It is therefore appropriate to look into the more modern data
storage devices now coming on the market. One of these is the
"videodisk", a device developed to store TV programs (1/2 hour per
disk), dictionaries, teaching aids, and in general for easy access
to large data banks. Such disks, which can be mailed and are
somewhat larger than a phonograph record with about the same thick-
ness, are now capable of storage of the order of 10^{10} bits, with
semirandom access. This number is equal to the number of bits
stored on 25 reels of 1600 bpi tape; storage of up to 500 digital
pictures of the type discussed above would be possible. They are
permanent, non-magnetic storage devices; the recording equipment
is expensive, but they can be reproduced extremely inexpensively
and the reading equipment is cheap.

However, the present development of mass-storage seems to go
in the direction of magnetic-tape type storage devices. In the
next five years, we may expect capacities of up to 10^{13} bits.
Ordinary tapes with a density of 6250 bpi have been in use for
some time. Digital recorders which can put 10^{11} bits on one reel
of wide tape (equivalent to 250 tapes of 1600 bpi density) are not
much more expensive than present standard tape units. Several
large computer centers are installing mass-storage devices which
can handle 10^{11} bits in a semirandom-access manner. It is to be
expected that most computer centers will have extensive mass-
storage capability well within a decade.

Suddenly, the data storage problem seems to have become considerably less severe. Moreover, having easy access to large data banks through the computer, regardless of the way in which these data were obtained, will lead to a very much improved interaction capability between the astronomer and his data. It is clear that the development of these devices has to be watched carefully, as the astronomer is of course wholly dependent on commercially available products in this area. But it should also be abundantly clear that international standardization becomes of the utmost importance if we are to set up easily accessible data banks. Not only should we decide on one type of device for use in astronomy everywhere, but we will also have to standardize the formats and even some of our reduction procedures to reach compatibility between data obtained at different observatories. Thus, with the availability of mass-storage devices around the corner, the main problem now is standardization of the data calibration, reduction techniques and presentation of "massaged" data.

POSITIONAL ASTRONOMY.

I will be brief on the area of positional astronomy. Undoubtedly modern electronic techniques will increase the accuracy and the amount of positional data, but in view of the nature of the reduction process (comparison of data, proper motions and precession corrections, etc.) which has to take place after the data are taken, efficiency of data recording at the telescope is only of importance to the investigator. However, it is becoming increasingly possible to digitize all plates, and in fact retain only the X and Y coordinates (and magnitudes) of all the stars on a plate. Plate constants can be as complicated as needed once the computer takes over. For proper motion studies, one might well want to compare such digital records, using basically all the stars on the plate. One might want to digitize archival plates. Perhaps one does not even want to keep the original plates, if all stellar data on these plates can be permanently stored on easily accessible mass-storage devices.

Once the positional data bank starts increasing by factors of ten, we should, as mentioned above, learn to live with compact data storage and the necessity for unpacking programs for every user of data tapes. The SAO catalog is a good example of a well-packed tape, but it takes a lot of computer time to decode the alphanumeric format if one wants to use the numbers directly in the computer. Yet even there it is easily possible to reduce the length of the tape by at least a factor of two, for example by recording RA and DEC in degrees and decimals, putting proper motion and its error in one integer word, etc., etc.

With computer controlled telescopes the necessity for large
data banks of accurate positional data will become increasingly
necessary. Modern telescopes have built into their steering
programs corrections for telescope flexure and refraction. This,
plus instantaneous measurement of and correction for the local
refraction (i.e. the deviation from the mean refraction) by means
of electronic guiding will make it possible to always set a telescope to an accuracy of one second of arc. This allows for
unambiguous identification of stars provided the star position is
known to that accuracy. Finding charts become in most instances
unnecessary. Some modern telescopes already have this capability,
and it seems imperative that every star to be investigated be
listed with its coordinates and, if available, proper motion
values, to be obtained from existing catalogs, measured from
plates, or measured during the initial survey from which the star
was chosen. The problem is again one of storage, but one million
positions (star number, RA, DEC, PM if available) can easily be
stored on one reel of 1600 bpi tape - and obviously much more
efficiently (and accessibly) on a mass-storage device.

PHOTOMETRY AND SPECTRAL CLASSIFICATION.

Finally, I would like to give another example of the possibilities inherent in having a computer on-line at the telescope.
I want to use this as an example of how one could possibly reduce
the data flow from telescope to investigator rather than an
example of how we can reduce our overall data storage problem.
Nowadays, a number of different devices are available to electron-
ically measure low-resolution spectra: spectrum scanners, multichannel devices, Fourier-transform spectrometers and adaptations
of array photometers. Let us assume that we have a device which
can produce a digital spectrum consisting of 1000 or more points
spaced over a relatively large range of wavelengths. This device
will continuously feed data into the computer which in turn can
display the spectrum on a CRT screen. The observer can watch the
spectrum build up on the screen, check for possible errors and
determine whether he has enough data (i.e. sufficient signal-to-
noise) for the problem under consideration. In the meantime, the
computer might do some checking of errors and interference itself,
for example by comparing every minute's worth of integration to
the first minute of integration, or to an average or a running
mean. If at the end of the observation we see the spectrum in
its full glory on the screen, what to do next? One should
realize that these electronic devices have a very high linearity
and of course can be calibrated extremely accurately. Therefore,
it is possible to correct the spectrum for gain as a function of
wavelength, provided the computer has remembered a previously
measured gain determination using a well-known standard lamp or
standard star or what-have-you, and apply corrections for zero-

level. These corrections can be made immediately after the
observer has decided he has a good enough spectrum. The next
step might be the correction for extinction. Storing well-
calibrated spectra of standard stars, it is of course possible
to determine instantaneous extinction values by looking at the
distribution of intensity with wavelength, i.e. using the differential extinction as a measure for the total extinction. In
addition, since a number of extinction stars will have been
measured throughout the night, the computer will be able to do
the normal type of total extinction measurement as well by integrating the total starlight over the spectrum - remember that the
spectrum is always calibrated on an absolute scale in some manner.

Now suppose one is interested in spectral classification of
such a star. It is possible to keep a large number of standard
spectra on the disk of the computer, and one can devise a program
where the computer quite accurately finds the spectral class of
the star that has just been measured (perhaps by giving the
computer a hint as to where approximately to search). Moreover,
the observer can see how good the fit is. It may turn out that
although the basic spectrum seems to fit there is a difficulty in
fitting the slope. Provided the extinction correction has been
applied properly, such a difference in slope would then obviously
be due to reddening. Presumably the computer will have a program
built in that takes out the reddening correction as well. Hence,
at the end of an observation a few minutes' fiddling with the
computer will produce a spectral type and a reddening value at
the same time. It may be that this information, i.e. two numbers,
is enough for the observer. Therefore a terrific economy in data
storage has been obtained, as the observer will only retain these
two quantities. However, the observer might wish to review his
data afterwards; in that case he will presumably put the entire
spectrum on tape for further work. Initially, in fact, he might
want to check out the programs by doing the entire reduction
procedure over with the large computer at his home institution.
He might, in fact, wish to put every one-minute integration on
tape, both of the standard stars and of the program stars.
Supposing he observes a 1000-point spectrum and puts it on tape
every minute, this will fill at most one-tenth of a reel of tape
per night. But it is obvious from the above that one might as
well do all reductions and come up with a final product immediately
after the observation. Alternatively, the observer might come
back before his next observing run and use the on-line computer
to read his tape, basically going over his observing run again
and doing the reductions over once more to make sure that his
interaction with the computer in the middle of the night was not
influenced by lack of sleep. It is also obvious that one can
play many other tricks in this way, such as having the computer
determine colors in one of the many color systems by simply applying the appropriate filter curves to the spectrum obtained. Here

again there is a distinct possibility for extra checks by comparing
the colors of standard stars so observed with the known colors of
these stars to make sure the program indeed works properly, and
calibrations have been successful.

In all of this, we have been talking about one on-line
computer. One might well consider having a microprocessor as a
data-routing device; it would collect and store the data, thus
leaving the minicomputer free for actual on-line data reduction,
perhaps while data on the next object is being collected. And a
warning is in order: quite a bit of programming effort is needed
to enable the observer to do on-line reductions; it is in general
not an easy and quick job that every observer can do for himself.

CONCLUSION.

In this paper I have only skimmed the surface. I did not
discuss high-resolution spectroscopy, classical photometry,
interferometry, space astronomy and many other areas where the
data flow is already large or is expected to increase considerably.
In some of the space astronomy centers, unreduced data has already
piled up in enormous quantities; a space telescope works 24 hours
per day and it is expensive to turn it off. I hope that my
examples have served the purpose of indicating my view that we
are on the brink of an entirely new era of data gathering and
storage technology. But in addition, the availability of data
in machine-readable form allows an entirely new process of interaction between the astronomer and his data. The astronomer used
to be part of the data processing and analysis; he was, one might
say, switched in series with the data. Data manipulation through
interactive use of the computer, where the astronomer is able to
look at his data selectively, massage them, look at subsets, try
out the results of different judgments, is the essence of this
new era. Minicomputers can now be equipped with disks containing
10^9 bits for a fraction of the cost of the computer; data
manipulation with direct access to a large data base is easily
possible even at relatively small institutions.

Finally, the very rapid development of minicomputers and
microprocessors necessitates a constant alertness to the availability of the many new possibilities for data handling and storage.
An example are the array processors, hard-wired devices able to
perform complicated but standard operations in parallel on very
large multidimensional arrays in extremely short time intervals.
A 1000-point Fourier transform is now possible in a matter of 10
microseconds. Another example is the use of such devices for the
implementation of efficient coding techniques. Storing a Fourier
transform of a data array rather than the array itself can be
extremely efficient if the array contains a large number of

blanks - as is often the case with astronomical photographs. The
possibilities are almost unlimited.

Summarizing, let me make the following four points:
1) Let us realize the present and future potential of the combination of minicomputer and electronic data gathering equipment. In
the latter, let us especially concentrate on the multichannel
capabilities, the inherent linearity and dynamic range, and the
possibility of instant calibration.
2) Let us realize that replacing the photographic plate by digital
devices will increase the data storage problem enormously and will
require the entirely new types of storage devices now becoming
available.
3) It becomes important at this time to start deciding which data
one actually wants to store for a long period of time. In the
last example, the question might well be asked: Do we want to
store the low-resolution spectra or are we satisfied with a properly classified spectral type and interstellar reddening? The
decision in the past was usually not very hard. A spectrum
appeared on a photographic plate (an extremely efficient storage
device) and plate vaults are filled with almost every conceivable
piece of information obtained at the observatories. But who uses
this data again? Is it useful to keep it? Clearly, we have to
judge internationally (through the IAU) what type of data we
really want to store for future use.
4) For the immediate future, while we are still using magnetic
tapes, we will have to design efficient data packing programs.
It is highly necessary that every data user be prepared to unpack
densely packed data in his home computer, while at the same time
data formats should be chosen such that unpacking programs are
easy to implement and require a minimum of computer time.

I have posed a number of questions, the main one being:
"What to do with the data flow to be expected?" I have not given
answers, other than pointing out the availability or need for
development of very compact data storage devices. What are we
going to do about it? I propose that a committee be appointed,
possibly under the auspices of the IAU or one or more of its
Commissions, to: a) prepare a report on the present use, and
present and future availability of data storage devices,
b) recommend certain types of storage devices for use in astronomy,
c) recommend standard formats for the various astronomical applications, and d) urge close collaboration between the various
institutions most actively engaged in the development of forefront
equipment. Perhaps a major effort of this type will lead to at
least a modicum of order in this rapidly developing field.

DATA PROCESSING AND ANALYSIS FOR SPACE-BASED ASTRONOMY

I. Mistrik

Max-Planck-Institut für Astronomie, Heidelberg, F.R.G.

ABSTRACT

The sensors on a spacecraft, the spacecraft data handling system, the mission support ground data system, the experimenter's scientific data reduction and analysis system, the post-mission data storage and dissemination system all comprise elements of the data handling system for a space-based astronomy.

Primary treatement is given to the experimenter's scientific data system where the following aspects of astronomy missions are discussed: increasingly close operational contact of the experimenters with the sensors calls for moving data processing operations to the earliest possible point of the system. On the other hand, potential need for subsequent reprocessing of data would favor the central ground-based facility for initial processing with the open-end system philosophy.

One approach, extant and planned implementation of the scientific data analysis system for the zodiacal light experiment of the solar probe Helios, is presented in detail, taking into consideration experience from the various astronomy missions, in particular Pioneer and Skylab.

INTRODUCTION

The general space-astronomy data system, depicted in Figure 1, consists of five major components, each element reflecting different physical location.

C. Jaschek and G. A. Wilkins (eds.), Compilation, Critical Evaluation, and Distribution of Stellar Data. 61-68.

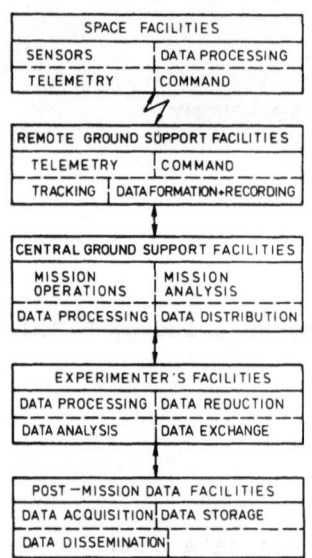

Figure 1 - General Space-Astronomy Data System

        Electrical signal from various sensors of the spacecraft
are initially conditioned and processed within individual experiment assemblies. The additional data processing may be done within
the spacecraft data handling system performing functions such as
raw data reduction without reducing information content, raw data
reduction which reduces information content, data gathering, data
compression or expansion, data storage, and data formatting and
coding. This is followed by the telemetry link, which includes
the transmitter, receiver, antennas and space path. The command
link allows modifications of the spacecraft configuration [1].

        The remote ground support facilities include the subsystems
required to establish contact with the space facilities and to
track their motions. Functions performed by these facilities include data acquisition, command, transmission, tracking and data
preparation for communication lines.

        The central ground support facilities perform the functions
of mission operations support, mission planning and analysis, distribution of various data records produced at these facilities.
There are three types of records coherently related to the preparation of permanent scientific data records: the system data
record (SDR), the master data record (MDR) and the experimenter
data record (EDR) [3]. The SDR is a log made at the central point
of the system for each of the network systems (Tracking, Command,

Telemetry, Monitor). The MDR has extraneous, and duplicate seg-
ments removed and the remainder is an organized, identified set
of records, in digital forms. The EDR contains information of one
particular experiment and supporting information such as orbit/
attitude-, housekeeping-, and command-data.

The EDRs are further processed by the experimenter's facilities, consisting of subsystems performing the following functions:
EDRs-processing, -monitoring, -displaying, and -converting, meta-
processing, several phases of data reduction, data analysis and
data preparation for scientific data exchange. The latter step
supplies the World Data Center with the analysis and physics tape.

The post-mission data facilities, on the world-basis referred
to as the World Data Center, disseminate data to the scientists
upon request. More specifically this center, functioning as the
integrated space science data center, is responsible for the active collection organization, storage announcement, retrieval,
dissemination, and exchange of space science data.

EXPERIMENTER SCIENTIFIC DATA SYSTEM

Figure 2 shows various levels of processing connected with
the experimenter's data system.

The straight forward operations level comprises various check-,

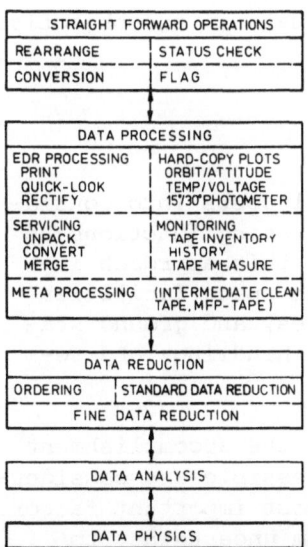

Figure 2 - Experimenter Data System-Processing Levels

flagging-, conversion-, and rearranging-procedures. There is very important limitation implying that only those procedures will be performed for which data can be acessed sequentially i.e. no tape movements backwards or forwards.

The data processing level consists of the following: EDRs processing, production of hard-copy plots for science-, engineering-, and orbital-data, monitoring of all processing, meta-processing (producing intermediate tapes necessary for further processing) and various converting, packing-, and merging-routines servicing the experimenter's data system.

Functional steps within the data reduction level are: ordering, standard data reduction and fine data reduction. The objective of the ordering step is to produce a tape with ordered (according to increasing measurement cycle number) data blocks. The standard data reduction step produces a reduced tape with preliminary measurement results (brightness, polarization and color). This step comprises straight forward procedures only, dealing with quantitative characteristics of data. The fine data reduction step consists of procedures requiring feed-backs of processed data. Qualitative characteristics of data are becoming more obvious. It produces an analysis tape with final measurement results (brightness, polarization and color).

The data analysis level comprises separation of components phase and averaging individual results phase. It produces a physics tape with final measurement results of zodiacal light measurements.

The theoretical interpretation of final data is the objective of the data physics level.

LOCATIONS FOR PROCESSING FUNCTIONS

Some arguments, taking mainly operational costs into consideration, call for the performing of data processing functions as close to the sensors as possible. Benefits of this approach include reduced communications bandwidth between space facilities and ground stations, ground stations themselves, and ground stations and central facilities; simplified data handling; and reduced logistics support.

On the other hand, arguments calling for the accomplishment of data processing as late in the system as possible are considering increased scientific effectiveness as a most important factor. This approach allows to improve implied system uncertainty,to change portions of the system and to exploit unexpected phenomena.

The appropriate approach should take into consideration the

both arguments. In general, the data processing for each of the operational functions mentioned earlier should be located according to the following criteria: degree of understanding of the system, state of knowledge of the physical phenomena being investigated, the experience of experimenters,and supporting personnel and operational costs. By evaluating criteria outlined earlier one should be able to choose the appropriate system in which the physical location for the data processing could eventually lean either toward spacecraft or toward experimenter's facilities. In many cases it should not be the closed-system. Basically, it should be possible to deliver essentially raw data to the experimenter if necessary to provide a high certainty that he can understand the data. Processing can later be moved progressively toward the spacecraft as that processing, the performance and the operations become more routine. Definately it should be possible to revert the earlier mode if circumstances require. It

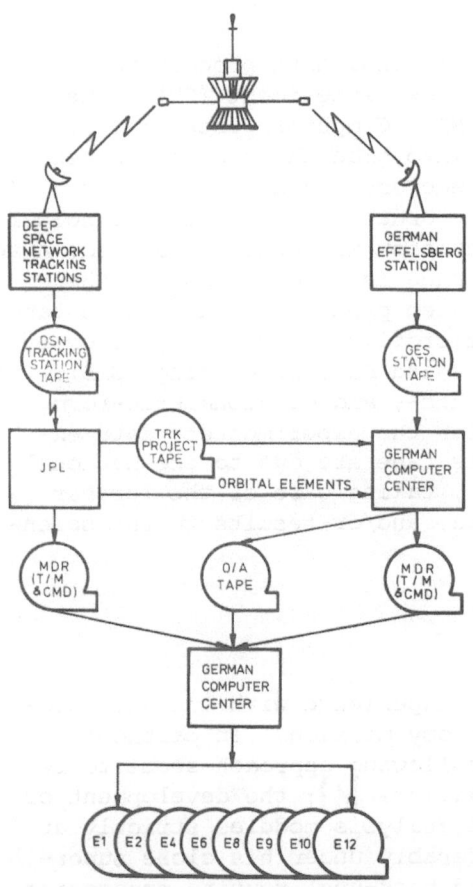

Figure 3 - Helios System Data Flow

is also desirable to retain flexibility to perform many functions
in more than one location to accomodate subsystems failures and
other unexpected situations.

OPERATIONAL SYSTEMS

The operational system, implemented for the german experimenters of the solar probe Helios, is shown in Figure 3 [2]. The
telemetry data are initially processed by the remote-processing
facilities (German and/or American stations) producing digital
tapes further processed by the central data processing facilities
(Jet Propulsion Laboratory (JPL) and German Computer Center).
These facilities are producing the following data records: tracking (TRK) tape, orbital elements, telemetry and command master
data records (T/M and CMD) tape and orbit/attitude (O/A) tape.
MDRs and O/A-tape are further processed by the german computer
center producing the experimenter data record (EDR) tape for each
experimenter.

Figure 4 illustrates the operational data processing and
analysis system for the zodiacal light experiment (E9) of Helios
probe. The Max-Planck-Institute (MPI) Computer Center merges
EDRs for the period of one month into ordered tape, having extraneous and duplicate elements removed, and data reorganized
into meaningful logical units. The ordered tape is the principal
input for the scientific data reduction and analysis system. After
the reduction phase the analysis tape is produced and further
processed by the data analysis system producing the physics tape.
This system is supported by the microfilm information system of
the German Society for Mathematics and Data Processing (GMD),
putting data on microfilm in graphic-, and/or alphameric-form.
The microfilm is primary medium for the experimenters data exchange. The reduced-, and analysis-tape are due to the National
Space Science Data Center (NSSDC), taking care of the further
dissemination of experimental-data, and/or-results to the scientific community.

CONCLUSION

With respect to the previous experience with the data processing for the space-based astronomy missions, in particular
Pioneer, Skylab and Helios, the following approach seems to be
favored by the principal investigatiors [4]: the development of
the scientific data reduction and analysis modules strictly at
the experimenter's facility, preferably under his close supervision; the centralized (with good back-ups) routine processing
and data handling; the possibility to use general software pack-
ages (e.g. integrated graphics system, file manipulation system)

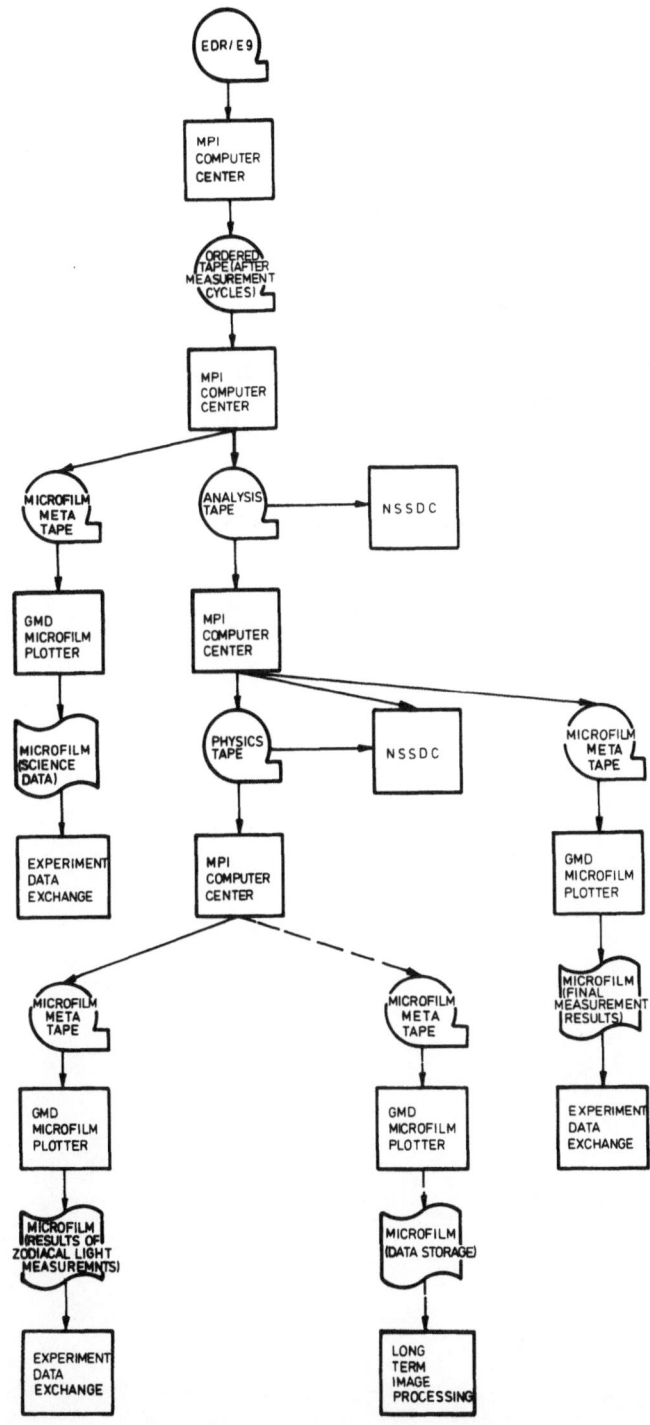

Figure 4 - Zodiacal Light Experiment System Data Flow

on service basis at the central processing facility (e.g. NASA-Center).

REFERENCES

1. Ludwig, G.H., Space Science Data Processing.
     NASA TN D-4508, May 1968.

2. Beard, E., Shout, C.M., GSFC Data Processing Plan for Helios-A.
     GSFC X-565-71-358, September 1971.

3. Mistrik, I., A survey of issues concerned with space science
     data processing.
     Raumfahrtforschung,
     vo. 18, no. 6, p. 263-267, December 1974.

4. Hanner, M.S., Leinert, C., Weinberg, J.L., 1976
     Private Communications.

THE USE OF STANDARDIZED DATA FORMATS WITH THE WESTERBORK RADIO
TELESCOPE

R.H.Harten and T.A.Th.Spoelstra

Netherlands Foundation for Radioastronomy,
Sterrewacht,
Leiden – 2405, The Netherlands

ABSTRACT: Since 1970 the Westerbork Synthesis Radio Telescope
has been collecting data on galactic and extragalactic fields.
The data is stored and cataloged on magnetic tape using a standard format. This data is routinely reduced and analyzed at
different observatories in Europe. A reduction program package
allows manipulation and display of the data. After the initial
reduction all tape output from the reduction programs uses the
same tape format regardless of the program. Transportation of
data between observatories is done via magnetic tape with the
same standard format. For data transportation to observatories
not normally using Westerbork information a common tape format
has been developed. Suggestions to alleviate problems with this
data transfer are discussed.

1. THE WSRT AS A SYSTEM

    Since 1970 the Westerbork Synthesis Radio Telescope (WSRT)
has been collecting data on a large number of galactic and extragalactic fields. It consists of an array of 12 equatorially
mounted 25 m paraboloidal dishes on an east-west line of which
the two eastern telescopes can be placed at any position on a
300 m rail track. The other ten telescopes are fixed and arranged
at equally spaced intervals of 144 m. In the second half of 1976
this system is enlarged with two additional telescopes, identical
to the existing ones, which will provisionally be added as move-
able telescopes to the existing rail track (Baars and Hooghoudt,
1974).

    At present correlations between each of the orthogonal di-

poles of a movable telescope and each of the orthogonal dipoles
of a fixed telescope form 80 interferometers at 20 different
baselines. When the two new telescopes come in operation, a new
5000 (frequency/polarization) channel auto-correlation receiver
will also be used. The observing wavelengths include 6, 21 and
49 cm. The 6 and 21 cm observations are done both in the continuum and line mode (Casse and Muller, 1974; Allen, Hamaker and
Wellington, 1974).

At present we have approximately 2×10^{10} words (64×10^{10} bits)
of data stored on magnetic tape. This data is divided into units
called maps ($\sim 2.5 \times 10^5$ words) which contain the brightness
distribution, at one of the three observing wavelengths, for a
region of the sky (typically $1^\circ \times 1^\circ$). Clearly there must exist a
well defined system to handle this volume of data. The basis for
WSRT reductions consist of a standard software package and a
fixed set of tape formats. Details of the reduction system have
been given by Van Someren Gréve (1974). The tape formats are
discussed below.

The processing of WSRT data can roughly be broken into four
stages: (1) observation and preliminary correction; (2) calibration; (3) Fourier transformation; (4) manipulation of map data
(display, etc.). Between each stage, data is written on magnetic
tape. The corrections in step 1 include instrumental corrections
(e.g. for delay, baseline, frequency, clock, gain, phase errors
and shadowing of the telescopes) and corrections for sky effects
(e.g. spherical refraction, precession, nutation, aberration and
ionospheric Faraday rotation). Other operations possible in this
stage are: shifting the field centre, corrections for objects
moving with respect to the fixed α, δ coordinates, subtraction of
point sources and additional gain corrections.

Reductions are considered to begin after the Fourier transform of the data when a map is made from the calibrated and
corrected interferometer output. Generally the maps are produced
in one of the four Stokes parameters. The output tape format for
this stage is used in all further reductions. All programs use
the same tape format for input and output. This holds for maps in
the v,l or v,m as well as l,m space (here v stands for velocity
in km/s; $l = \sin (\alpha - \alpha_0) \cos \delta$, $m = -\cos (\alpha - \alpha_0) \cos \delta$ and
$\alpha - \alpha_0$ is the right ascension with respect to the field centre).
The expanded WSRT system will require a 3-dimensional Fourier
transform (l,m,v) but the data will remain stored in a 2-dimen-
sional form (see below), but with a 3-dimensional cross-reference
index. The reduction programs will accept both the old and new
formats.

This data is routinely reduced and analyzed at at least four
different observatories in Europe. There exists a standard

reduction program package which also allows manipulation and display of the data. The manipulation includes map combination, projections, polarization parameter determination, cleaning of a map (Högbom, 1974), etc. The display modes include line plots, contour plots and optical intensity modulated photos. Also interactive programs have been developed for manipulation of these data at various observatories.

2. TAPE FORMATS

There are four tape formats used in data processing for the WSRT: (a) a format for Telescope tapes on which the observations are transported from the telescope to the computing centre where off-line corrections and calibrations are done, (b) a format for Makeobs tapes containing output from these corrections and calibrations, (c) a format for the reduction stage (Makemap tapes, i.e. Linemap tapes for the new system), (d) a format for tapes with which the results are transported to other institutes (Transport tapes, see Sections 3 and 4). The first is used solely by the reduction group and the users generally work only with the last three tape formats. The formats are summarized in Table 1.

3. TRANSFER OF DATA VIA TAPE TO OTHER INSTITUTIONS

Several other observatories use Westerbork data on a regular basis. Previously, data had been transfered via punched cards. This method has proved to be unsatisfactory. The chance for punch errors is too high and for large volumes of data it is impractical. To solve this problem, we developed a special tape format which was readable on almost any computer. This allows us to send data without special programs for each computer model. Also, users could use our tapes even if they switched computers. We write the data and catalogs in BCD card image format in records of 120 bytes long. Several of these records can be packed into a block to save space on the tape. Generally we write 7-track even parity tapes. However we also use 9-track odd parity if the institution is very modern and doesn't have the older 7-track units. To prevent problems of label reading, we do not write standard tape labels, only data and tape marks. The record size of 120 bytes is quite important since it is divisable by 4, 6, 8, 12 or 60. These are the most common word sizes in various computers presently in use. The block size should be an even number of 120 byte records. For many users the 240 byte length is quite handy since it represents 3 punched card images. Clearly, this format is not designed for maximum efficient use of the space available on a magnetic tape, but it reduces the amount of word needed to read and write transport tapes.

TABLE 1

Tape	Area	Format [1]	Length [2]	Contents	Organization
A Telescope	Start Catalog	Mixed: 2-4 б I, 4-8 б R, 1-4 б L, 1 б A	116	Parameters for object identification, receiver parameters and corrections added to the observation	1 dataset per label (each dataset consists of 1 start catalog, ≥1 data blocks and 1 end catalog)
B	Data	(as for entry A)	636	Interferometer output per baseline per time interval	
C	End Catalog	(as for entry A)	116	(as for entry A)	
D Makeobs	Catalog	Mixed: 2-4 б I, 4-8 б R, 1 б L, 1 б A	160	Parameters for object identification, observational identification	≥1 dataset per label (each dataset consists of 80 sets of 1 catalog and 1 or 2 data blocks)
E	Data	2 б I	≤4096	Corrected observations recorded per time interval per baseline	
F Makemap	Catalog	Mixed: 2-4 б F, 1,2,4 б I	4096	Information to define the map and its parameters; description for each baseline whose data was used in making the map	1 dataset per label (the number of blocks is determined by the number of points in a map. Any unfilled part of the last block contains meaningless values). The basic unit is a map or antenna pattern and its catalog

TABLE 1 (continued)

Tape	Area	Format	Length	Contents	Organization
G	Data 3)	4 Ƀ F	4096	Values of Brightness/Velocity/Polarization values on an evenly spaced grid	
H Linemap	Master Catalog	(as for entry F)	3840 4)	Information common to an entire set of maps: i.e. coordinates, map size & cross-reference of all maps in the set	
I	Map Catalog	(as for entry F)	1920 4)	Information about a single map: i.e. frequency, bandwidth, etc.	The number of blocks is adjustable. It is possible that a second block is needed
J	Data	Mixed: 2-4 Ƀ I	30720 or 3840 5)	See entry G. The data in this section are the mantissa of a floating point number. All numbers have some exponent which is stored in the catalog. The first bit is a sign bit, the remaining are the mantissa.	Data and catalogs are organized to allow for a 3-D array of data. For each set of 2-D maps which make up the 3-D array, there is a master catalog. Each map consists of a map catalog and a data section. Map catalog and master catalog are also written at the end of the label.

Notes to Table 1:
1) 1 Ƀ = 1 byte = 8 bits
   I = integer; R = real;
   L = logical; A = alphabetic;
   F = floating point number
2) in bytes
3) Map and Antenna Pattern have the same format
4) 120 Ƀ records
5) 3840 Ƀ for CDC machines

4. STANDARD INDENTIFICATION BLOCK

At present when data is transfered between observatories, either a copy of an existing tape is made or a special tape is created. Usually, the contents of the tape are described (in varying degrees of detail) in a separate description. In many cases the description is lost or not available and the user has to go to some effort to find out what is on the tape and how it is written. Even if the tape is well documented, some of the technical details such as parity, block length, etc. are not described. To help alleviate this problem, we suggest that each tape begin with an identification area. This area would contain a description of the tape format and its contents. The area would have a fixed, standard block length but the number of blocks would be adjustable. The number of blocks in the area would be written in the first block. The information written in the area would be in BCD card image format. Thus the contents of the area could be read by a simple FORTRAN, ALGOL or PL/1 instruction. To minimize problems we suggest that the tape be written without standard labels. Thus, the tape would consist of the identification area, tape mark and as many data areas, followed by tape marks, as needed. The information contained in the area would be written in a form that is

TABLE 2

Identification Area

A. Number of records/blocks of 240 bytes.
B. General Format of Tape:
   Number of labels, labeling conventions used, parity, BCD, EBDIC or binary. Machine used to create the tape.
C. General Description of the data on the tape and its purpose including:
   Type of data on the tape i.e. radio, infrared, spectral data, etc., format of data (grid of some quantity on a grid of some coordinates, including description of coordinates), intensity units, and any general notes pertaining to all data on the tape.
For each label there should be an entry describing:
D. Technical Parameters of data in the label:
   Block length of the data in the label; number of blocks in the label; format of data within the label.
E. Description of contents or type of information written on tape:
   Description of field coordinates, grid size and astronomical source. Calibration procedure and sources used; telescope, beamsize (or grating), wavelength, filters used, bandwidth, integration time, noise level, person with knowledge of observing procedure and calibration.

---

self defined, thus no description is necessary. A suggestion for
the contents of such an area is outlined in Table 2. While the
proposed identification area may not be the optimum solution to
the problem, we suggest that some form of standard identification
be written on all tapes used to transport information between
observatories. The value of the system is that a tape's contents
can be identified without a description and by someone not
familiar with the tape.

The record/block length for the identification area would be
240 bytes per record/block. This would equal three card images
and would be handleable on most computers. The first record would
contain the number of records/blocks in the area. This would be
a 4 digit number at the start of the record/block. The rest of
the information would be in BCD character form. The identification
area would be followed by a tape mark. Unfortunately, one must
know the parity and number of tracks (7 or 9) of the tape before
it can be read. If a standard is adopted 7 track, even parity or
9 track, odd parity are reasonable possibilities.

REFERENCES

Allen, R.J., Hamaker, J.P., Wellington, K.J., 1974, Astron.&
                  Astrophys. 31, 71
Baars, J.W.M., Hooghoudt, B.G., 1974, Astron.& Astrophys. 31, 323
Casse, J.L., Muller, C.A., 1974, Astron.& Astrophys. 31, 333
Högbom, J.A., 1974, Astron.& Astrophys.Suppl.Series 15, 417
Van Someren Gréve, H.W., 1974, Astron.& Astrophys.Suppl.Series
                  15, 343

DATA STORAGE REQUIREMENTS IN RELATION TO RADIO INTERFEROMETRY
OBSERVATIONS

H. G. Walter

Astronomisches Rechen-Institut,Heidelberg, F.R.Germany

Summary

Preliminary results of a search for observational data of
radio sources are presented. Emphasis is put on those items which
are imperative to make such a data collection amenable to astrometrical exploitation with a view to the identification of optical
and radio positions, and the selection of suitable radio stars as
reference points for other galactic and extragalactic objects.

1. Introduction

In the past years radio interferometry observations of radio
stars and compact extragalactic radio sources yielded positional
data which are competitive with meridian circle observations.
Apart from minor exceptions radio star positions have been derived
from observations by means of short baseline interferometry (SBI),
while positions of extragalactic radio sources were obtained by
SBI (Elsmore and Ryle, 1976; Brosche et al., 1973) as well as by
very long baseline interferometry or VLBI (Rogers et al., 1973;
Cohen, 1972).

From this background the existing fundamental astrometric
system receives new impulses (e.g. Fricke, 1972), and the role of
radio interferometry in the redetermination of astronomical constants has been recognized. Before, however, extragalactic radio
sources become eligible as representatives of an extragalactic
reference frame which ought to be formally related to the conventional stellar reference system as closely as possible some consistent efforts in acquiring and treating of observational data
are indispensable.

2. Celestial Radio Sources as Reference Points

It is practical to distinguish between radio stars of the Galaxy and compact extragalactic radio sources, the latter ones being comprised of two types: (1), radio sources emitting in the radio frequency domain only and, (2), radio sources with optical associates. While the first type is simply and solely capable to establish a radio astronomical reference system independent of the adopted stellar system, the second type possesses also the potentiality of tying the radio system into the stellar system under the hypothesis of coinciding optical and radio centres of the sources (Fricke, 1974; Walter, 1976). On comparing optical and radio positions of appropriate extragalactic objects the disclosure of inhomogeneities across the sky and, perhaps, their elimination is rendered possible.

Other than extragalactic radio sources the radio stars are affected by proper motions and, hence, are no ideal reference points. Nevertheless they are invaluable intermediaries between the highly accurate positions of extragalactic radio sources and the stars of the fundamental stellar reference system, provided sufficient radio stars are tied to it. Furthermore, they are instrumental in supplying the radio observations with the zero point of right ascension. Unless either appropriate radio sources in the planetary system or precise lunar occultations of radio sources are available, the radio astronomical reference system suffers from the deficiency of letting the vernal equinox undetermined. - At all events, referencing of extragalactic radio sources to the conventional stellar system is one of the presuppositions for further refined studies on kinematics and dynamics of the Galaxy.

3. Extragalactic Radio Sources

To date radio interferometric position measurements of some astrometrical bearing are found in nine lists which are compiled in Table 1. Five of them comprise positions of extragalactic radio sources obtained by means of short baseline interferometry and the remaining ones contain positions derived from VLBI observations.

Radio positions are dependent on the fixing of the origin of right ascension. Since the observing groups adhere to different conventions, their individual results impede an immediate comparison, but suggest to have recourse to FK4 as an expedient. Pursuing along this line Moran (1975) arrived at standard deviations in right ascension and declination of the order of 0.1 arc sec. Besides, the comparison was impaired by the small number of sources common to the lists and by processing measurements acquired over a wide range of frequencies, because different frequencies may refer to different locations of the emission centres. Furthermore, interferometry over baselines several thousand kilometers long tends to resolve the

Establish-ment	Technique	Length of Baseline [km]	Freq. [MHz]	Number of Sources	$\sigma(\alpha)$ $\sigma(\delta)$	Comments	References
RRE, Malvern	SBI, phase	0.7	2695	159	$0\overset{s}{.}03$ $0\overset{''}{.}4$	relat.posit.	Adgie et al.,1972
NRAO,Green Bank	SBI, phase	2.7	2695 8085	59	$0\overset{s}{.}03$ $0\overset{''}{.}5$	rel.RA abs.DECL	Brosche et al.,1973
WSRT,Wester-bork,Holl.	SBI, phase	1.6	1415	233	$0\overset{s}{.}07$ $1''$	rel.RA rel.DECL	Katgert, 1973
REE, Malvern	SBI, phase	0.7	2695	161	$0\overset{s}{.}03$ $0\overset{''}{.}5$	relat.posit.	Adgie,1974
Mullard, Cambr.,Engl.	SBI, phase	5	5000	53	$0\overset{s}{.}008$ $0\overset{''}{.}03$	rel.RA abs.DECL	Elsmore et al.,1976
Cal.Tech., Pasadena	VLBI, fringe rate	~4000	7850	11	$0\overset{s}{.}013$ $0\overset{''}{.}2$	rel.RA rel.DECL	Cohen,1972
MIT,Cambr., Mass.	VLBI,del.and fringe rate	845 3930 3325 ~4000	7900	12	$0\overset{s}{.}006$ $0\overset{''}{.}1$	rel.RA quasi abs.DECL	Rogers et al.,1973
MIT,Cambr., Mass.	VLBI,del.and fringe rate	845 3930 3325	1660 7900	8	$0\overset{s}{.}01$ $0\overset{''}{.}2$	rel.RA abs.DECL	Whitney,1974
NRAO,Green Bank	VLBI, fringe rate	3325	1428	13	$0\overset{s}{.}03$ $0\overset{''}{.}3$	rel.RA abs.DECL	Hemenway,1974

Table 1. Review on radio interferometric position measurements of extragalactic radio sources

radio sources and, as a consequence, the generated signal exhibits
much complexity leading in certain cases to ambiguous positions.
This effect is rather widespread since, virtually, most extragalactic radio sources show some angular structure.

Once the accuracy of the VLBI technique has been fully realized the evaluation of the observational data with a view to precise
position determination must account for the dependence of the
measurements on frequency and baseline length.

4. Radio Stars

On searching the literature for galactic radio sources about
seventy of these objects were found and for 52 of them the radio
emission is attributed to a star. Their positions are plotted in
Fig.1. The distribution of the objects is far from being uniform;
they accumulate distinctly in the direction to Cygnus and to the
galactic centre. For about 25 of them the places have been determined by radio interferometric measurements to an accuracy of
better than one tenth of a second of arc in both coordinates on
the average.

A classification of radio stars by apparent magnitude, radio
flux and frequency is attempted in Table 2. The majority of the
observations have been made with the NRAO interferometer of 2.7 km
baseline length; a few observations each were provided by the 5 km
Cambridge interferometer and by the Westerbork interferometer.

Numerous radio stars produce only low radio fluxes, thus
complicating precise position determination, especially when, in
addition, binary stars are involved and radio emitting and optical
components urge on proper identification. Further aggravation is
caused by the flare stars which at times are bright according to
radio astronomical standards but are characterized by variable
fluxes.

Although the present overall situation is not yet quite ideal
for imparting to radio stars the role of reference objects, it will
steadily improve when radio astronomers succeed in making repeated
observations of the same radio stars at different epochs under
identical instrumental conditions, and in decently distinguishing
between absolute and relative position determinations in the specific sense of radio interferometry.

5. In Conclusion

The positional data obtained so far through radio interferometry demonstrate the feasibility of setting up a reference system
of extragalactic radio sources and relating it to the stellar
system. Before radio sources really qualify as basic constituents

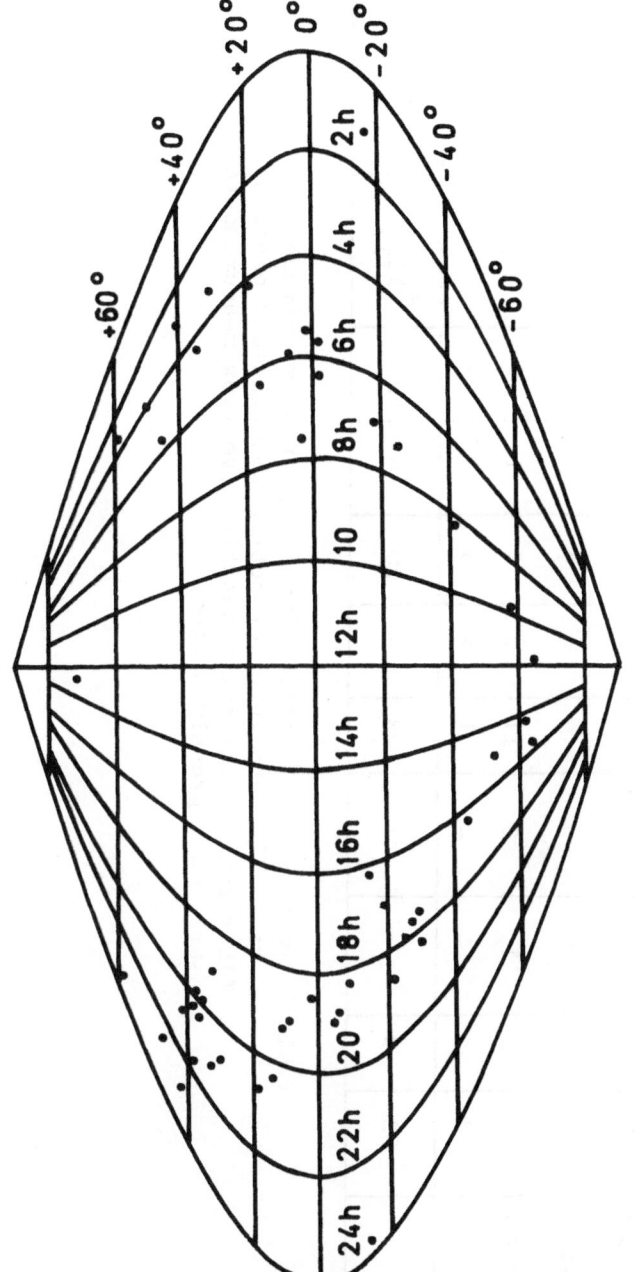

Fig.1. Distribution of radio stars: northern and southern hemisphere with 29 and 23 stars respectively.

	No.of Radio Stars	Radio Flux S $[10^{-29} Wm^{-2} Hz^{-1}]$ *)				No.of interferom. obs.of radio stars MHz				FK4 st. AGK3 st. SAOC st.			Flare stars Norm.single st. Binary stars		
		<10	10..20	20..50	>50	1415	2695	5000	8085	FK4	AGK3	SAOC	Flare	Norm.single	Binary
m ≤ 7.0	9	4	3	1	1		5	2	5	5	6	8	0	2	4
7.0 < m ≤ 9.0	10	2	3	1	4	1	4		5		7	8			4
9.0 < m ≤ 11.0	7	1			6		3		3			1	1		1
11.0 < m ≤ 14.0	23	2	1	3	11	2	5	1	5				3		
14.0 < m	3			2	1	4	3	3	3						
Total	52	9	7	7	23	7	20	6	21	5	13	17	4	2	9

Table 2. Radio star classification by apparent magnitude, radio flux and interferometric observation techniques. *)For some of the stars quantitative flux measurements are not expressly stated.

of the radio astronomical reference system dozens of absolute interferometric observations of some tens of extragalactic radio sources uniformly distributed over the sky are required with several years' of epoch spacing. Uncertainties in determining positions from VLBI measurements are conveniently avoided if the baseline geometry and the technical features of the interferometers are kept constant to a great extent.

Acknowledgement

In searching the literature for radio stars the author could avail of the catalogue of radio emission of stars by Wendker (1975).

Literature

Adgie, R.L., J.H. Crowther, H. Gent, 1972, Mon.N.R.astr.Soc., 159, pp.233-251

Adgie, R.L., 1974, Astron.J.79, pp.846-851

Brosche, P., C.M. Wade and R.M. Hjellming, 1973, Aph.J., 183, pp.805-818

Cohen, M.H., 1972, Astrophys.Letters 12, pp.81-85

Elsmore, B. and M. Ryle, 1976, Mon.N.R.astr.Soc., 174, pp.411-423

Fricke, W., 1972, Annual Rev.of Astron.Astrophys.10, pp.101-128

Fricke, W., 1974, Astronomisches Rechen-Institut, Heidelberg, Mitteilungen Serie A, No.89

Hemenway, P., 1974, Ph.D. Thesis, University of Virginia, Maryland

Katgert, P., J.K. Katgert-Merkelijn, R.S. Le Poole, H.van der Laan, 1973, Astron.Astrophys.23, pp.171-194

Moran, J.M., 1975, Space Research XV, Akademie-Verlag Berlin, pp.33-47

Rogers, A.E.E., C. Counselman III, H.F. Hinteregger, C.A. Knight, D.S. Robertson, I.I. Shapiro, A.R. Whitney, T.A. Clark, 1973, Aph.J., 186, pp.801-806

Walter, H.G., 1976, Mitteilungen der Astron.Ges.Nr.38, pp.185-188

Wendker, H.J., 1975, Catalogue of Radio Stars, Hamburger Sternwarte, Hamburg

Whitney, A.R., 1974, Ph.D. Thesis, MIT, Cambridge, Mass.

ULTRAVIOLET PHOTOMETRY FROM THE S2/68 OBSERVATIONS IN THE TD1
SATELLITE

K. Nandy

Royal Observatory, Edinburgh

SUMMARY

   An ultraviolet photometric system based on observations
obtained from the ultraviolet Sky Survey Telescope in the TD1
satellite is described. The system considered here consists of
ultraviolet magnitudes at λ_1 = 2740A, λ_2 = 2190A and λ_3 = 1490A.
The extinction free parameters derived from the observed ultra-
violet colours enable determination of interstellar reddening,
spectral type and luminosity. This photometric system has been
evaluated by comparing these parameters with other classification
parameters.

   The observations from the S2/68 Ultraviolet Sky Survey Telescope in the ESRO satellite TD1 have provided homogenous data on
stellar spectra in absolute fluxes in the wavelength range from
2500A to 1350A over an area covering approximately 80% of the sky;
the spectral resolution is $\sim$ 30A (Boksenberg et al., 1973). In
addition the experiment gives a broad band measurement centred at
2740A of width 310A at half maximum. The number of observations
per star is usually 3, this number being greater for stars near
the ecliptic pole. The stars brighter than V=9^m at B0 to 8^m at A0
have been detected and identified. The estimated number of stars
observed is over 40000. Since the number of observations does not
depend on the brightness of the star, the photometric accuracy of
the spectral data points becomes low for stars fainter than V=6^m.
For brighter stars with small amount of reddening the accuracy of
the mean spectrum is better than 3%. These spectra are individu-
ally useful and an ultraviolet catalogue of bright stars is going

to be published. A number of distinct features appear in these
spectra, the strength of which vary with spectral type and
luminosity (Cucchiaro et al., 1976). For the fainter stars,
however, the spectral data points have to be combined into wider
wavelength bands to give statistically useful results. The narrow
band ($\Delta\lambda \sim 100A$) magnitudes have been constructed at several wavelengths from the spectral data, and the intrinsic ultraviolet
colours, their dependence on spectral type and luminosity and the
effect of interstellar reddening have been studied from the extensive data available from the S2/68 experiment (Nandy et al., 1976a).
On the basis of these results an ultraviolet photometric system
has been considered, which determines spectral type, luminosity
and reddening (Nandy et al., 1976b). In this paper this photometric system will be evaluated.

Because of the complexity of the spectra in the ultraviolet
wavelength region due to the presence of many absorption lines,
the colour indices need to be selected with considerable care for
an ultraviolet photometric system. The wavelengths have been
chosen (1) to avoid the strong spectral features, (2) to avoid
the extremities of the spectrophotometer channels and (3) to
include the wavelength at which maximum interstellar extinction
occurs. The wavelengths chosen are $\lambda_1 = 2740A$, $\lambda_2 = 2190A$ and
$\lambda_3 = 1490A$, λ_2 being close to the wavelength of the ultraviolet
extinction peak. λ_2 and λ_3 are chosen in view of the observations
that the luminous stars have flux deficiency increasing with $1/\lambda$
as compared to the main sequence stars of similar spectral types
(Humphries et al., 1975).

The fluxes obtained at these wavelengths are converted to
magnitudes m_λ, where $m_\lambda = -2.5 \log I_\lambda - 21.1$ (Oke and Schild,
1970). The photometric accuracy of m_λ has been discussed by Nandy
et al. (1976a). The ultraviolet colour $(m_{2740} - V)$ primarily
determines the spectral type (colour temperature) while the colour
$(m_{1490} - V)$ is sensitive to both spectral type and luminosity
(Nandy et al., 1976b). The two extinction free parameters have
been derived from the observed colours, using $(m_{2190} - V)$ as a
reddening indicator. These parameters are defined as follows:-

$$\text{PHI} = (m_{2740} - V) - \frac{E_{2740} - V}{E_{2190} - 2740} (m_{2190} - m_{2740});$$

$$\text{and} \quad \text{PSI} = (m_{1490} - V) - \frac{E_{1490} - V}{E_{2190} - V} (m_{2190} - V).$$

The mean values of the colour-excess ratios (see Nandy et al.,
1976a) are:

$$\frac{E_{2740} - V}{E_{2190} - 2740} = 1.1; \quad \frac{E_{1490} - V}{E_{2190} - V} = 0.8$$

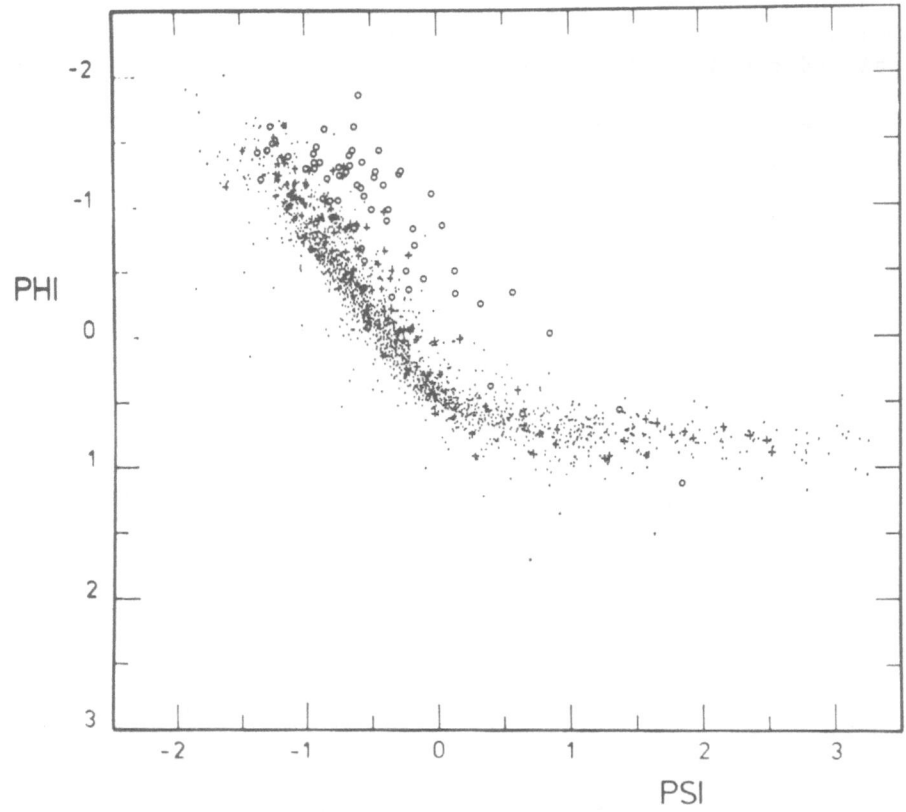

FIGURE 1. The relation between PHI and PSI. Circles denote class
    I and II stars, crosses class III and dots main sequence
    stars.

The analysis of the extensive stellar data obtained from the S2/68
experiment shows that the extinction laws derived for different
galactic regions are not significantly different. All the observations have been used to give a single mean extinction curve.
However, there are a number of individual cases which exhibit significant deviations from the mean extinction law; these deviations
may arise due to the presence of circumstellar or local clouds
close to the stars (see, for example, Willis and Wilson, 1976).

    Fig. 1 shows the relation between PHI and PSI for several
thousands stars. Except for a few stars, the points fall naturally
into two groups. The points denoted by dots lie in a fairly
narrow region in the lower part of the diagram and all of these
are class V. Also many points lie significantly above (open circles) and these belong to class I and II. The class III stars
denoted by crosses tend to lie between these two sequences. The

importance of the PHI - PSI diagram is that the stars can be classified according to their positions in this diagram. Also, any
star which shows an anomalous extinction law due to circumstellar
or localised cloud can be identified.

The correlation between PHI and Q, and PSI and Q are shown
in Fig. 2, where

Q = (U-B) - 0.72 (B-V).

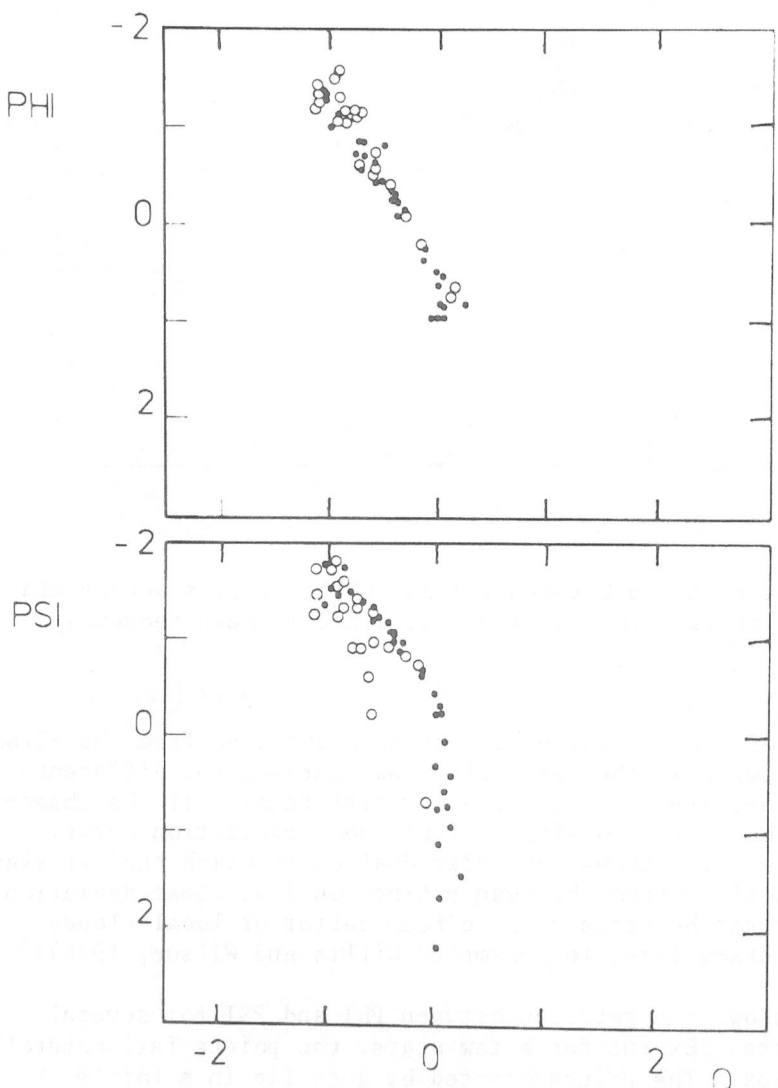

FIGURE 2. Relation between PHI and Q (top) and between PSI and Q
(bottom). Symbols as in Fig. 1.

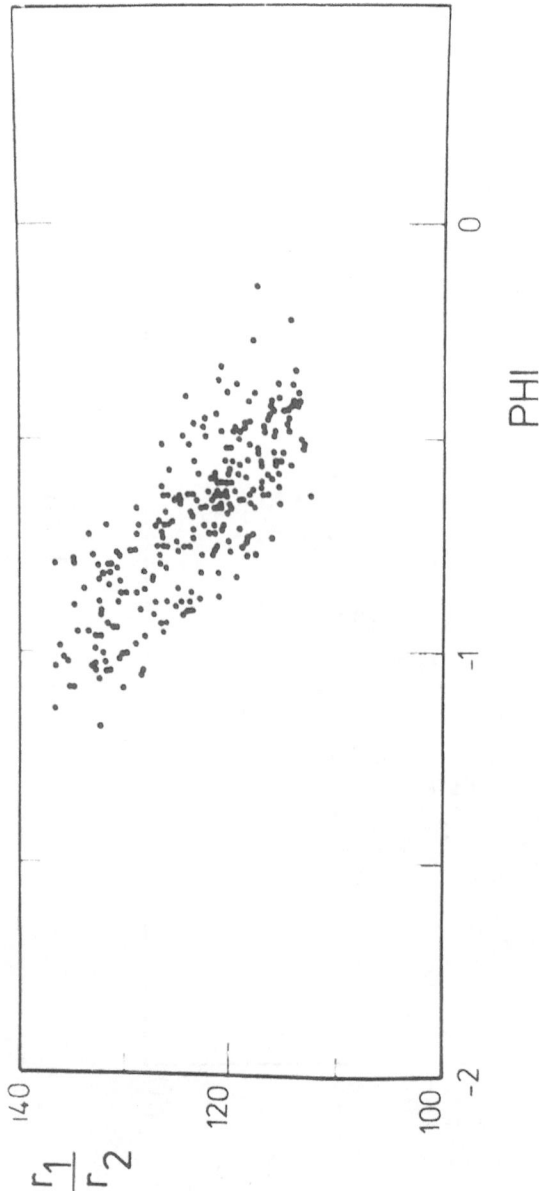

FIGURE 3. Relation between r_1/r_2 and PHI (see text).

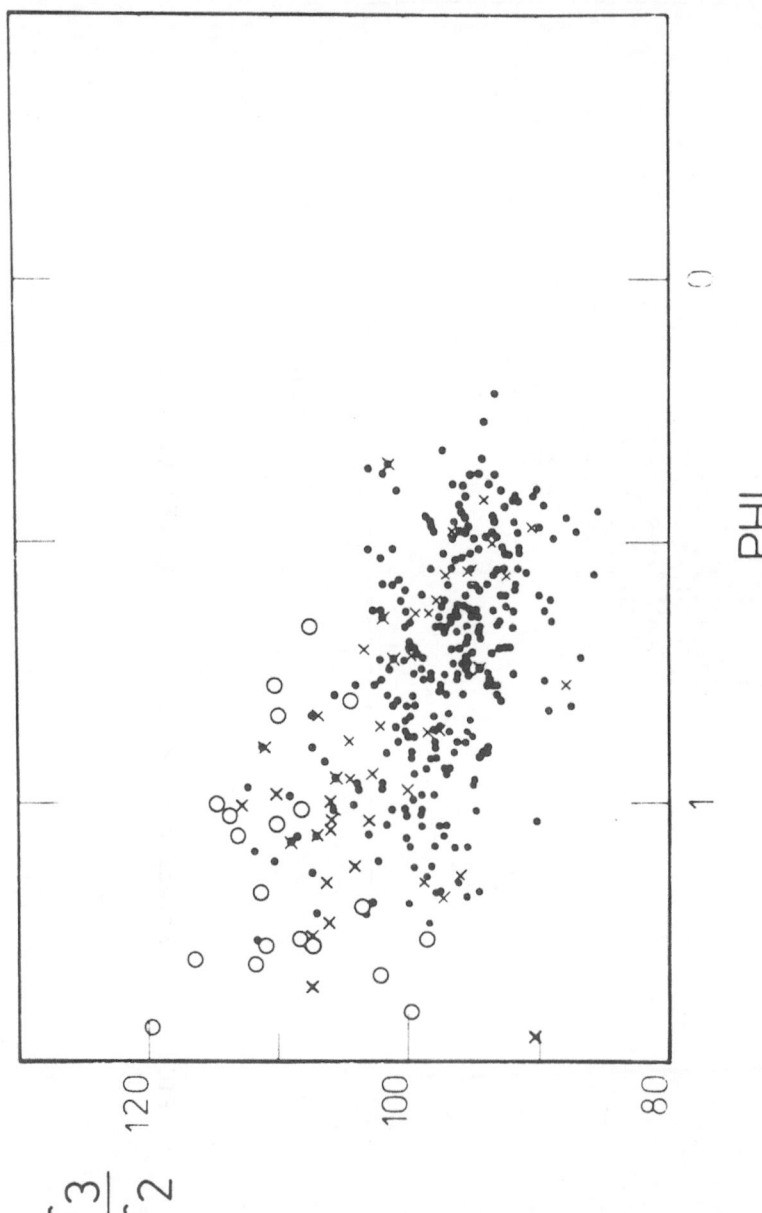

FIGURE 4. Relation between r_3/r_2 and PSI (see text). Symbols as in Fig. 1.

The mean relation between PHI and Q is linear, and there is no separation between dwarfs and luminous stars, whereas in the PSI vs Q diagram the supergiants are clearly separated from the main sequence stars.

Cucchiaro et al. (1976) have used the intensity ratio of the features occurring at 1410A and 1550A (denoted by r_1/r_2) and of the 1620A and 1550A features (denoted by r_3/r_2) for a two dimensional classification. The intensity ratios are independent of interstellar reddening. The relation between PHI and these intensity ratios are shown in Figs. 3 and 4.

It is shown that the classification parameters derived from the observed ultraviolet colours are well correlated with the parameters derived from the intensity ratios of the spectral features. Therefore for the fainter stars the ultraviolet colours $(m_{2740} - V)$, $(m_{2190} - V)$ and $(m_{1490} - V)$ allow separation of dwarfs from luminous stars and determination of interstellar reddening and spectral type. This photometric system will provide homogenous classification of a large number of stars distributed all over the sky. This work is in progress.

REFERENCES

Boksenberg, A., Evans, R.G., Fowler, R.G., Gardner, I.S.K., Houziaux, L., Humphries, C.M., Jamar, C., Macau, D., Macau, J.P., Malaise, D., Monfils, A., Nandy, K., Thompson, G.I., Wilson, R., Wroe, H. 1973, Mon. Not. R. astr. Soc., 163, 291.
Cucchiaro, A., Jaschek, M., Jaschek, C. 1976, Astron. and Astrophys. (in press).
Humphries, C.M., Nandy, K., Kontizas, E. 1975, Astrophys. J., 195, 111.
Nandy, K., Jamar, C., Monfils, A., Thompson, G.I., Wilson, R. 1976a, Astron. and Astrophys. 51, 63.
Nandy, K., Humphries, C.M., Thompson, G.I. 1976b, I.A.U. Symposium No. 72 (Lausanne).
Oke, J.B., and Schild, R.E. 1970, Astrophys. J., 161, 1015.
Willis, A.J., and Wilson, R. 1976, Astron. and Astrophys., 44, 205.

THE UNIVERSITY OF TEXAS CATALOG OF ULTRAVIOLET AND OPTICAL STELLAR DATA

Sidney B. Parsons and James D. Wray

Department of Astronomy, University of Texas, Austin, U.S.A.

At the University of Texas we are working on the reduction and analysis of ultraviolet stellar spectra obtained with the S-019 experiment on <u>Skylab</u> (Karl Henize, principal investigator). Our objective prism photographs recorded nearly 10,000 stars at wavelengths shortward of 3000Å, many of them of 9th or 10th visual mag., and for proper analysis we need to assemble ground-based data into a workable master catalog. The most immediate application of this catalog will be computer simulation of the ultraviolet photographs in order to point out significant anomalies with respect to previous knowledge. We need to have a catalog, then, which is as complete as possible in terms of the existence of early-type stars to at least 10th mag., their spectral types, photometry, known peculiarities, duplicity, and so forth. Also the master catalog must contain only a single value for each magnitude and for the spectral type, rather than a listing of various determinations, in order to be useful for the simulation. This means that auxiliary data files are needed for recording data in more detail so that the master catalog may be kept at a workable size, while at the same time the basic data are readily available.

Another use we intend for this catalog is as a reference data base to which we can add quantitative measurements of the ultraviolet spectra in order to produce a convenient catalog of S-019 ultraviolet observations. Figure 1 shows examples of scans of widened

C. Jaschek and G. A. Wilkins (eds.), Compilation, Critical Evaluation, and Distribution of Stellar Data. 93-101.
Copyright © 1977 by D. Reidel Publishing Company, Dordrecht-Holland. All Rights Reserved.

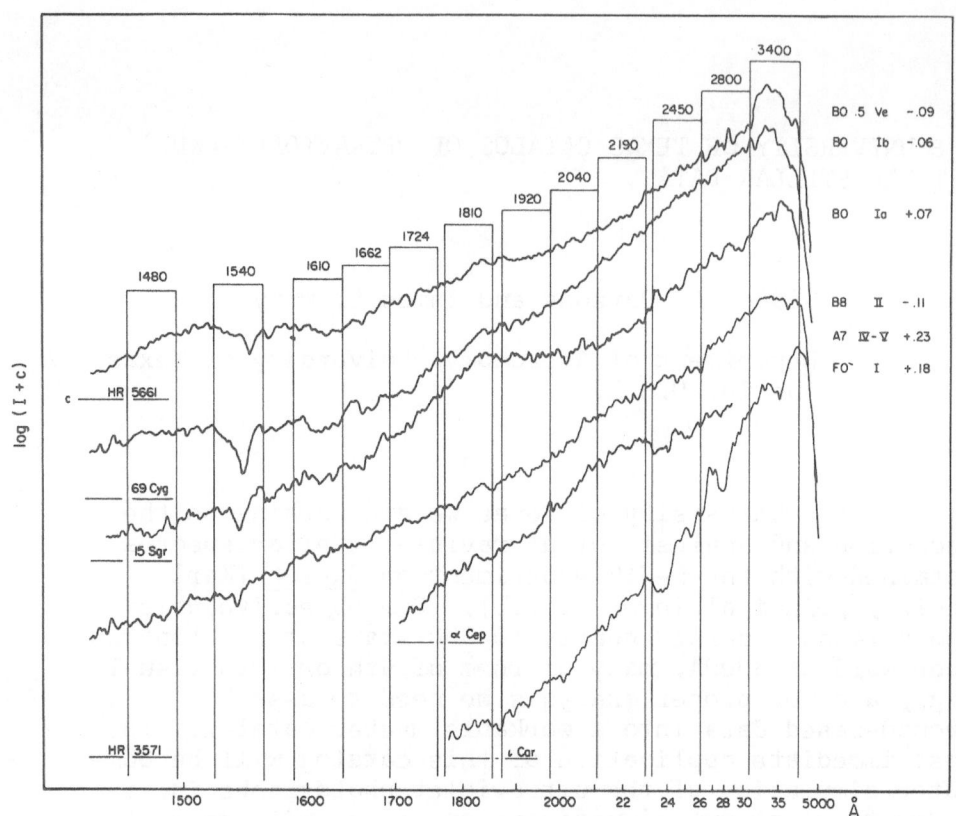

Fig. 1 -- Intensity scans of some S-019 objective prism
spectra, showing proposed system of pass-band measurements.
B-V color index is given after MK spectral type for each
star.

spectra and a proposed system of up to 12 intermediate
band measurements on each star. This is what we see
as the optimum way to present the ultraviolet fluxes for
these thousands of stars. We have scanned a few hundred
of the widened spectra so far parallel to the dispersion,
but the reduction to intensity is complex. For the
unwidened spectra, it will be best to scan perpendicular
to the dispersion. Such measurements on spectra which
extend at least to 2100Å will allow clear separation
of temperature from interstellar reddening, and we should
be able to determine the spectral class of any early-
type star to within 2 subclasses using only these measurements.

With our 4 x 5° field of view, we covered about 9%
of the sky during the manned Skylab missions. This

included about 25% of the area within ±15° of the galactic
equator. Our current work in developing a master catalog
at a low level of funding is therefore restricted to the
observed areas of sky in order to make the task manageable.
Our starting point is the Celescope Identification Catalog
which was assembled by the Celescope project prior to
launch of the OAO-2 satellite. The CID contains data
from 1000 references for 100,000 objects, although many
of these are straight from the SAO Star Catalog. Material
from the literature has been condensed in the CID into
just a few data items - magnitudes, spectral type, and
up to 21 characters representing different codes for
spectral peculiarities, variability, duplicity, and the
existence of certain data such as polarimetry and interstellar line measurements. Non-stellar objects were
included from the NGC and IC catalogues and also from the
literature. Although many checks were performed during
the condensation process, the result suffered from the
enormity of the undertaking. For example, it appears
that HD types were averaged in with MK types. Also, stars
in clusters were punched as non-stellar objects, often
resulting in duplicate catalog entries. HD numbers were
apparently deleted. Even so, this catalog is still the
best one available for our purposes, but it needs a
great deal of editing to make it useable.

We are now in the process of comparing and merging
several other catalogs with the Celescope data base.
Figure 2 outlines some of the characteristics of our
major source catalogs. Our main needs are to have
magnitudes, colors, spectral types, 1950 positions, and
references to spectroscopic studies. We hope to draw
on several other catalogs in the near future, particularly
the compilation of best spectral types by Mercedes
Jaschek, the Bibliographical Star Index, and the Michigan
Spectral Catalogue. We have actually edited and merged
only 7 declination strips so far, but this has included
much software development over the last 5 months.

Figure 3 shows a few entries from our master catalog
in the -24° declination strip. The University's CDC
6600/6400 system software uses 60-bit words, with 6 bits
per character for alphanumeric data. It is easiest to
work in terms of card image format, so that one line of
80 characters is 8 computer words. Thus we have set up
the catalog as shown for the easiest manipulation and
editing. The editing routine which we use, developed
mainly by a student, Mr. Dwight West, searches most
rapidly on any string of characters in the first word of

	Entries Year	Coords.	m_v, m_{pg}	UBV	HD type	MK type	literature references
Celescope Identification Cat. C I D	100,000 1969	1950	−	+	−	+	✱
Celescope Ultraviolet Cat. C UV	5,000 1973	1950, 2000	−	✱ incl. Suppl.	−	+	✱
Cat. of Stellar Identifications C S I	403,000 1974	1950	+	●	+	●	●
Photoelectric Cat. (Blanco et al.) PEB	33,000 1968	1900, 1950	●	✱	−	+	✱
Cat. of UBV Phot. (Mermilliod) PEM	29,000 1974	●	●	✱	●	+	✱

✱ = always present;   + = present if available;   − = present if no better data available;   ● = not present

Fig. 2 -- Summary of data available from several major catalogs on magnetic tape.

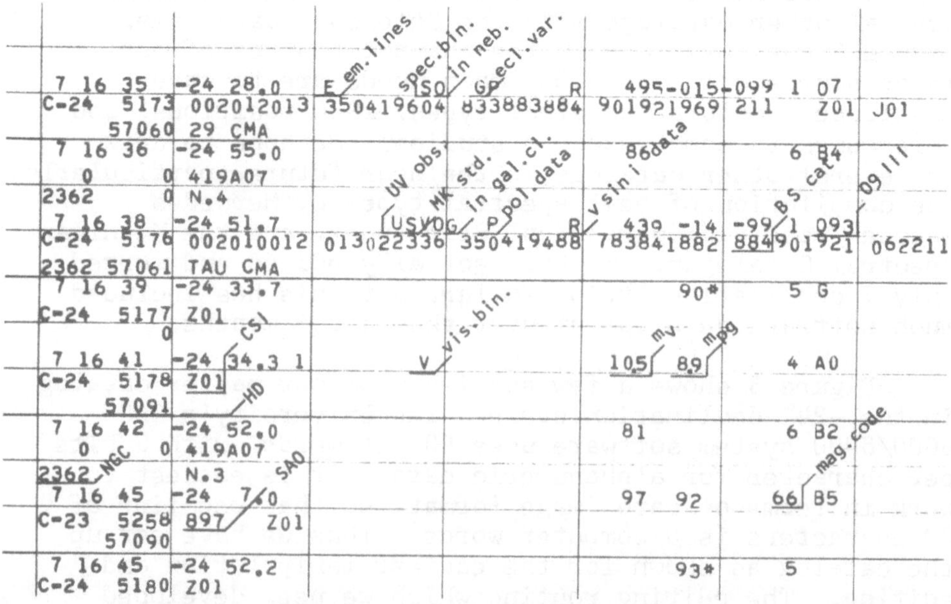

Fig. 3 -- Sample entries from master catalog. Three lines are used for each star, starting with 1950 R.A. and dec.

a line; thus we have started the 3 lines per star with the 1950 right ascension, the DM number, and the HD and/or NGC number. An HD star can be located by keying in a one-character command and the HD number, but one has to know or at least guess which declination file to call up. Some examples of the peculiarity codes and notes are shown here.

There is room for three magnitude values, with a magnitude code to tell the user what they represent. For example, a 1 means V, B-V, and U-B values, while a 4 means m_V and m_{pg}. The declination word has room for designating the component of a multiple star system. The 3-character spectral and luminosity classes are taken directly from the CID, while more detailed representations are being added in words 7 and 8 as they are found in the photoelectric catalogues. We intend to keep the 3-character representation as the working value, modified when appropriate, while words 7 and 8 will be used to show the range in MK classification. Most of the second line is available for a string of 3-character reference codes. The 2nd and 3rd words of the third line are used for remarks, such as a Flamsteed designation or alternate DM number. This latter is entered automatically if available from the CSI and if the star has no HD number. The rest of the 3rd line is available for future expansion.

Figure 4 gives a few examples of references and their codes which are being added to our data base. The format is quite similar to that of Celescope. We are trying to have somewhat mnemonic codes, e.g. C for classification and G for studies of specific groups or clusters. We are adding to this bibliographical data file as we find relevant articles in the literature, although we will have to complete the basic work on the other declination strips before data from the articles can be effectively entered into the master catalog.

Our procedure for handling the merger of the several catalogs is significantly affected by the advantages and limitations of the available computer system. We have to use tapes much of the time, although it is costly. The maximum disk file we can work with is 64000 words (8000 card images), while the practical limit for efficient turn-around time with the interactive editing routine is about 2000 card images or less than 700 stars. So far our declination strips, with stars excluded if outside of our observed sky or if much fainter than our limiting

AAB. DEUTSCHMAN, W.A., DAVIS, R.J., AND SCHILD, R.E.   1976
AAB. THE GALACTIC DISTRIBUTION OF INTERSTELLAR ABSORPTION AS DETERMINED
AAB. FROM THE CELESCOPE CATALOG OF ULTRAVIOLET STELLAR OBSERVATIONS AND
AAB. A NEW CATALOG OF UBV, H-BETA PHOTOELECTRIC OBSERVATIONS.
AAB. ASTROPHYS. J. SUPPL. 30, PP.97-225.

C02. WALBORN, N.R.   1973
C02. THE SPACE DISTRIBUTION OF THE O STARS IN THE SOLAR NEIGHBORHOOD.
C02. ASTR. J. 78, PP.1067-1073.

C03. BIDELMAN, W.P., AND MACCONNELL, D.J.   1973
C03. THE BRIGHTER STARS OF ASTROPHYSICAL INTEREST IN THE SOUTHERN SKY.
C03. ASTR. J. 78, PP.687-733.

D21. GREENSTEIN, J.L.   1974
D21. A NEW LIST OF 52 DEGENERATE STARS. VII. ASTROPHYS. J. 189, PP.
D21. L131-L133.

G02. HILTNER, W.A., MORGAN, W.W., AND NEFF, J.S.   1965
G02. STUDIES IN SPECTRAL CLASSIFICATION. II. THE H-R DIAGRAM OF NGC
G02. 6530. ASTROPHYS. J. 141, PP.183-186.

J01. MERMILLIOD, J.-CL.   1973
J01. A CATALOGUE OF UBV PHOTOELECTRIC PHOTOMETRY. C.D.S. INF. BULL.
J01. (STRASBOURG), NO. 4, PP.20-21 (1974 TAPE EDITION).

Fig. 4 -- Sample of references in bibliographical data
file, which adds to the references given in the Celescope
Catalog of Ultraviolet Stellar Observations.

magnitudes, are within the working limits. But our
procedures would be much simpler if bulk storage were
available so that random access from several catalogs
could be carried out at the same time.

The next two figures outline our present operations.
Figure 5 shows that we begin by rewriting the given tapes
so that unneeded stars are eliminated and the formats
are compatible. The condensed CID tape has one file
for each 1950 declination strip, so a few files at a time
are transferred to disk storage. Before we can do any
merging, a CID file must be compared with the CSI in
order to check the coordinates and the compatibility of
other data for the star, and obtain the HD number if
there is one. To accomplish this, the condensed CSI
tape, arranged by DM number, is searched over the 3
relevant DM zones for stars having the right declination.
These then have to be sorted by R.A. and stored on disk.
An interactive program operated at a CRT terminal searches
that file for all entries within 1.5 of a given CID entry
and displays them. If the coordinates are within 0.1,
the DM numbers identical, and the magnitudes and spectral
class similar, the program automatically inserts the HD
number in the CID. Otherwise it waits for judgement by
the operator. The resulting CID file can then be merged
with the relevant stars from the CSI. In this batch job,

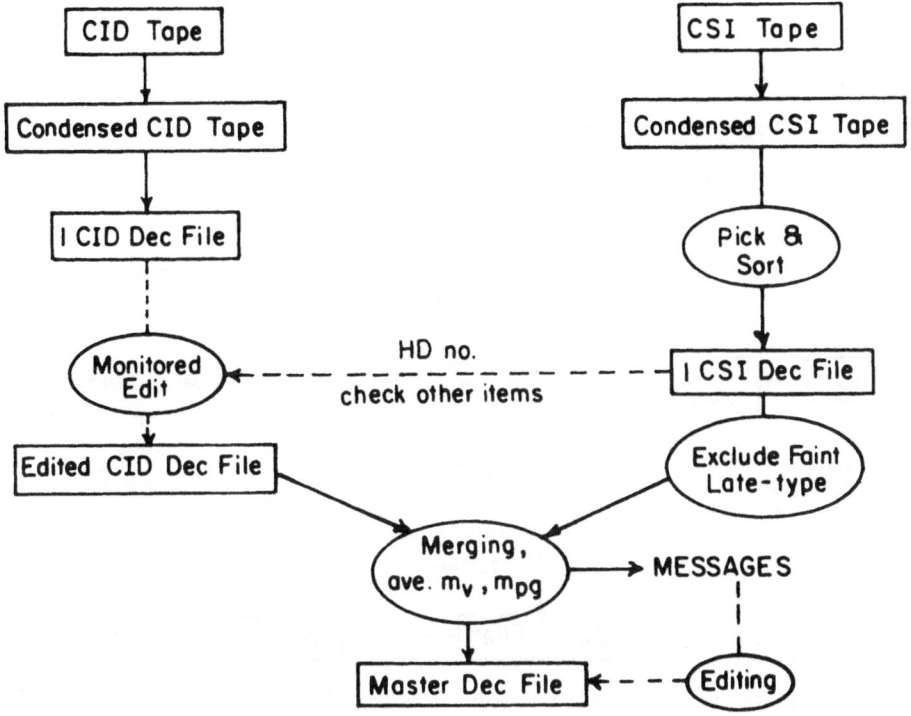

Fig. 5 -- Flowchart for the creation of a master file
for each declination strip, following a rewrite of the
CID and CSI tapes. Solid lines represent batch processes
while dashed line represent interactive processes.

tight requirements are again applied in testing for
matching stars, and a message is printed in case of an
inconsistency or of a multiple star.

Figure 6 shows schematically what happens next.
Information from other catalogs is entered into this
declination file mainly in batch jobs which print out
messages whenever individual attention and possible
subsequent editing are called for. These jobs require
first that the catalog tapes be rewritten into compatible
format for comparison with the master catalog. For
individual journal articles, it will depend on the nature
of the contents whether it is better to call up the
appropriate master declination file to enter new information, or to enter data into buffer disk files which
are arranged by HD or DM number. These files act as
a free-format repository of miscellaneous data, which can
later be used to create or modify entries in the master
catalog.

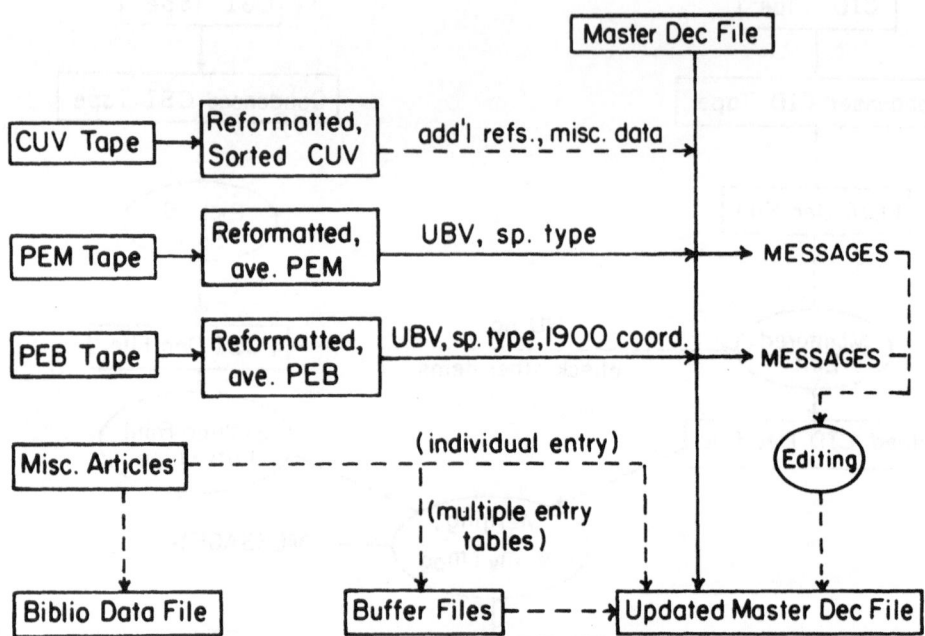

Fig. 6 -- Schematic representation of addition of data to the master file for each declination strip. Solid and dashed lines represent batch and interactive processes.

We urge that in the publication of data, astronomers bear in mind two things: 1) Be sure the title of the paper has a high information content: useful searches can be performed on titles if they are specific enough. 2) Strive for redundancy in the identifications of objects and give at least approximate coordinates for any non-DM stars. An HD number is normally sufficient identification, but one or more digits can get changed during the publication process. We suggest that the DM number always be given for an HD star; in addition to the useful redundancy, it will allow us to locate the star easily in our system of declination strips. Definitely refrain from listing a star only by an identification number referring to a chart in an obscure publication.

The buffer files are similar to W. P. Bidelman's card file of stellar data, although his is arranged strictly by 1900 coordinates. His file covers especially the early work on stellar spectra; he searched major journals up to 1950 and most observatory publications up through 1961; data from selected articles since then have also been entered. We purchased a bulky Xerox copy of his card file and began to punch the data, as shown in Figure 7. But our funding for the Skylab data analysis has dropped to the point where we cannot continue this

```
6100 +1128                    11.0 P              +11 1084  LS VI +11 2
6100 +1128 09 H
6100 -1930          MA  7.9              43236    -19 1391
6100 -1930 M3 III*APJ112*48 / *CIR
6101 +2854          A2P 7.3              43246    +28 1062  AS(MWC) 122
6101 +2854 F0*APJ81*1A7 (MV2*1) / COMP.*B1 / *JRASC42*140 / *APJ112*72 / *AJ52*
6101 +2854 128(2 SP) / PV VAR.*APJ88*50 /*HA56*82 /*APJ SUPP1*216(N*7) / *DAO13
6101 +2854 *119(ORR)
6101 +2250                                         N. 172
6101 +2250 08*T*T BOL*N*13*6
6101 +0707                    9.0 I
6101 +0707 C*APJ125*195


R.A. (1900) DEC       (1950)    VIS  OTHER  M(MIN)                 NAMES          SP(HD)
6 10.0 +11 28.   6 12.8  11 27    V 11.0 P      HD      DM +11 1084  LS VI +11 2
  08 H
6 10.0 -19 30.   6 12.2 -19 30    7.9 V         HD43236  DM -19 1391                  MA
  M3 III         APJ  112*48                                        CIR
6 10.1 +28 54.   6 13.3  28 53    7.3 V         HD43246  DM +28 1062  AS(MWC) 122     A2P
  F0             APJ  81*187  (MV2*1)                    COMP.       B1
                 JRASC 42*140                                       APJ  112*72
                 AJ  52*128  (2 SP)                      RV VAR.     APJ  88*50
                 HA  56*82                                           APJ SUPP 1*216   (N.7)
                 DAO  13*119  (ORB)
6 10.1 +22 50.   6 13.1  22 49    V             HD      DM           N. 172
  08             T*T BOL  *N*13*6
6 10.1 + 7  7.   6 12.8   7  6    V  9.0 I      HD      DM
  C              APJ  125*195
```

Fig. 7 -- Sample of cards punched from Bidelman's stellar data file. Upper section shows listing of actual cards; lower section shows printout from a program which checks the formatting and precesses coordinates to 1950.

project. The first line per star contains items for its identification, while succeeding cards contain data and references in free format. The lower section shows printout generated from the cards by a proof-reading program. We have applied for funding from the N.S.F. to complete the punching and prepare a tape for distribution. It will take two years of steady half-time punching to finish the job, but it will be an excellent complementary file to the Bibliographical Star Index, the CSI, and the catalogs of spectral classifications.

We hope that these efforts will lead to a more general data base in the next few years. We are optimistic about surveying the whole sky in the ultraviolet from a payload on the Shuttle spacecraft. Such a survey will demand a highly useable and comprehensive data base for its interpretation if a large scientific return is to be expected within a few years of the observations.

This work was supported by NASA under contract NAS 8-31459.

PART III

THE CRITICAL EVALUATION OF DATA

THE CRITICAL EVALUATION OF STELLAR DATA

Anne B. Underhill and Jaylee M. Mead

Goddard Space Flight Center

Theresa A. Nagy

Computer Sciences Corporation

ABSTRACT

Many catalogues of astronomical data appear in book form as well as in a machine-readable format. The latter form is popular because of the convenience of handling large bodies of data by machine and because it is an efficient way in which to transmit and make accessible data in books which are now out of print or very difficult to obtain. Some new catalogues are prepared entirely in a machine-readable form and the book form, if it exists at all, is of secondary importance for the preservation of the data.

In this paper comments are given about the importance of prefaces for transmitting the results of a critical evaluation of a body of data and it is noted that it is essential that this type of documentation be transferred with any machine-readable catalogue. The types of error sometimes encountered in handling machine-readable catalogues are noted. The procedures followed in developing the Goddard Cross Index of eleven star catalogues are outlined as one example of how star catalogues can be compared using computers. The classical approach to evaluating data critically is reviewed and the types of question one should ask and answer for particular types of data are listed. Finally, a specific application of these precepts to the problem of line identifications is given.

C. Jaschek and G. A. Wilkins (eds.), Compilation, Critical Evaluation, and Distribution of Stellar Data. 105-119.
Copyright © 1977 by D. Reidel Publishing Company, Dordrecht-Holland. All Rights Reserved.

I. INTRODUCTION

In recent years there has been a trend away from publishing catalogues in book form to preparing catalogues by computer and distributing them on magnetic tape with appearance in book form a secondary occurrence. For instance, Kelly and Palumbo (1973) assembled and cross-checked their compilation of atomic and ionic wavelengths shortward of 2000Å using punched cards. They then output the table on magnetic tape and finally prepared a tape to run a linotype machine for preparing the book. At present we have in machine-readable form catalogues that were prepared using a computer and catalogues which first appeared in book form and later were transcribed to a machine-readable format. A need exists for critical evaluation of all this data in order to find its machine-readable characteristics as well as the scientific validity of the data itself.

This need raises the following questions: How does one evaluate data and transfer the evaluation with the data? This concerns the documentation accompanying the data file. What is the best way to express an evaluation? Does one do this by means of a weighting system or does one prepare a written evaluation? What properties of a data file should be evaluated? Are there standard tests which should be applied and for what properties of files of astronomical data?

There are two types of catalogue: (i) a listing of data obtained by one method of measurement, for instance, a radial-velocity catalogue prepared from observations made at one observatory, and (ii) a compilation of a selected type of data from many sources, for instance, the U.S. Naval Observatory catalogue of UBV photoelectric photometry (Blanco, et al. 1968). If one is to use several machine-readable catalogues efficiently, one needs not only a cross index giving the ID's of the astronomical objects contained in the catalogues but also a preface describing what each item in the catalogue is and how it was obtained. Also one needs to know what accuracy may be expected, what systematic and random errors occur, and the completeness of the data.

In every catalogue selection rules have been applied. Thus, the Smithsonian Astrophysical Observatory Star Catalogue (Whipple 1966) lists only 4 to 6 bright stars per square degree. Not all bright stars in a crowded region are given. Other catalogues such as the radial-velocity or photometric catalogues prepared at certain observatories give data for stars of selected spectral types, within specified magnitude limits and within definite declination limits.

In evaluating a catalogue one needs to know what data are given and their sources as well as the selection rules which have been applied in making up the list of objects treated. When printed catalogues were the only sort available it was easy to obtain this

type of information. It was usually printed as a preface to the
catalogue. With machine-readable catalogues one does not always
have the needed information. It should be mandatory to provide
written documentation with each catalogue tape describing what was
in the original preface in the case of old catalogues and documenting fully new catalogues which have been prepared entirely
by machine. Many old catalogues are out of print, yet the data
contained in them remain valid. These data frequently are now
made accessible by means of machine-readable catalogues. A
determined effort should be made to develop and distribute
documentation for these machine-readable catalogues that preserves
the information given in the original preface.

II. CATALOGUE PREFACES

A preface should be prepared and distributed by those who
compile the catalogue or who distribute machine-readable copies of
an old catalogue. This preface should:

(1) describe the observational data used to provide the tabulated
    characteristic,

(2) give a detailed description of the instrumentation used to
    obtain the data,

(3) describe the methods used to obtain the tabulated character-
    istic from the raw measurements,

(4) describe the selection rules used to define the group of
    objects studied, e.g. area of sky covered, magnitude limit,
    spectral distribution considered, atomic species studied, etc.,

(5) give sources for the material used when the results from a
    series of catalogues or papers have been collated,

(6) describe the weighting system used, if any, to express an
    evaluation of the data,

(7) describe the search for and evaluation of any systematic trends
    in the data,

(8) give a study of the random errors in the compilation,

(9) describe the statistics of the objects reported in the cata-
    logue.

With many old catalogues no longer in print, it becomes
urgent to reproduce in printed form the essential information
given in the old prefaces. Should such information be published
as one or more articles in a scientific journal or should it be
published as a special publication of a government institution?
Whatever happens, the information will slowly be lost. One might
consider whether it is appropriate to transfer this needed descriptive material on a magnetic tape as a header record to the
original catalogue.

III. HANDLING ASTRONOMICAL CATALOGUES ON MAGNETIC TAPE

A general utility program can be used with a magnetic tape
of a catalogue to dump the contents of the tape or to scan them.
These initial reads will determine the properties of the data
control block, the number of tracks on the tape, and the type of
character code used. In addition they will count the number of
physical records and the number of files on the tape. If
documentation is available, this information can be verified; if
it is not available, it may be possible to generate appropriate
documentation by comparing with the original source.

The following procedure is followed when checking out an
astronomical catalogue on magnetic tape in the Laboratory for
Optical Astronomy at the Goddard Space Flight Center. First one
logical record from each block is listed using a high-speed input/
output routine. Also the last record is listed to ensure that
the entire catalogue is present. This information serves as an
index to the entire tape. Next a printout of the first one or two
hundred sequential records is obtained to checkout the corres-
pondence of the machine-readable version with the printed version
of the tape and/or accompanying documentation. A visual scan of
the printout is made to verify that given columns of information
line up for subsequent records.

Next a check is made on quantities which should be increasing
or decreasing, for instance catalogue identifier or right
ascension. Then a check is made on columns which are typically
numerical in nature, for instance, magnitude, to make sure that
all entires are indeed numerical. Sometimes if a value is not
available or is off scale, a substitute, such as asterisks, is
given. Since reading these symbols with a numerical format will
cause an error, the field is then changed to some large but
readily distinguishable number, for instance, 99.9, so that a
numerical format read will be valid.

A check on ranges of values to point out gross keypunching
errors is made, such as:

$$0 \le \alpha^h \le 23$$
$$0 \le \alpha^m \le 59$$
$$0 \le \alpha^s \le 59$$
$$-90 \le \delta^o \le +90$$
$$0 \le \delta' \le 59$$
$$0 \le \delta'' \le 59$$
$$\sim -5.0 \le m_V \le \sim 20.0$$

spectral types
luminosity ranges
galactic coordinates

If the catalogue which is being machine-checked has other identifiers, then a cross check with each identifier can be made to verify any common information. By this cross-checking technique, errors in either or both catalogues can be found. The greater the number of catalogues that are available (a definite advantage of a data center) the greater will be the reduction in the number of errors remaining in any given catalogue.

Minimum problems between computers are realized when a tape is written in either EBCDIC or external BCD. Otherwise, special techniques need to be employed which can be very time consuming. For instance, the IBM 360 series cannot represent in binary a -0, for zero will always be given with a + sign. If the record is written in EBCDIC, however, logical tests can be made to differentiate between the two cases.

Computer installations vary in their ability to handle magnetic tapes. Some can process only 7 or only 9 track tapes; not all density ranges can always be handled even if the computer can accommodate both track sizes. Some computers do not have FORTRAN-callable routines to process multifile tapes.

Older machine-readable catalogues often tried to limit information to that which could be carried on an 80-column computer card. In order to accomplish this, they would use overpunches in a given column or columns to represent different cases. Efforts to unravel these overpunches can be quite time consuming especially if the overpunches are not well documented.

In an effort to decrease the number of blank spaces in a given record, frequently only a small number of columns are allotted for a given class of information. An example of this is other catalogue identifiers. It would be more convenient for the user of each catalogue to have each major identifier in its own set of columns. One must keep in mind that the reason for having machine-readable catalogues is so that the machine can easily retrieve any given set of information.

IV. THE GODDARD CROSS INDEX

With the multitude of catalogues that exists in machine-readable form, it is necessary that a cross index exist relating the different identification numbers (ID's) of a star or other astronomical object to each other. At the Goddard Space Flight Center we have developed a cross index that accesses eleven star catalogues. It is the result of merging these star catalogues using the Mark IV File Management System. To use this technique, one first designs a framework into which each catalogue is arranged to fit. For each stellar entry, there is provision for right ascension, declination, visual magnitude, photographic magnitude, spectral type, proper motion and the ID from each catalogue to be merged. Not every one of these "spaces", or

quantities, will be filled for every star, of course. The
computer program is tailored to process each catalogue in order to
copy and rearrange the catalogue data to fit this master plan or
layout. We refer to any catalogue rearranged in this way as a
Submaster.

For most catalogues, the common or linking ID is either the
Henry Draper (HD) number or the Durchmusterung (DM) number. We
have divided each catalogue according to these two basic ID's:
stars having HD number; stars not having HD number, but which
do have DM number; those having neither. The HD stars were
then sorted by increasing HD, the DM stars were ordered by
decreasing DM zones starting at the North Celestial Pole. Those
stars having neither HD nor DM number have not yet been merged
into the Cross Index.

Once each catalogue has been prepared to fit the required
format, it must be ordered in the same sequence as all the other
catalogues to be merged, and precessed, if necessary, to the same
epoch. One is then ready to add the catalogues, or Submasters,
one at a time, to the Master Cross Index. This step consists of
matching and merging the Master and a Submaster, where there is a
common ID of either HD number or DM number. If the computer finds
a match between two HD's, it then tests the positions given by
the two catalogues. If the two positions are separated by more
than 0.1 degree of arc, the ID from the Submaster is not entered
into the Master. Instead, the entire record for that star, as
given in the Submaster, is copied into an error file and the
Master for that entry is left unaltered. The stars in the error
file are later hand-checked against the original catalogues and
the key-punched version of the catalogue in an effort to uncover
the cause of the discrepancy in position. Sometimes the HD
number for one star has been punched incorrectly, and the computer
attempts to treat two <u>different</u> stars as if they were the same
since their HD numbers read the same. Sometimes a sign error in
the declination or a punching error in the position is discovered in either the Submaster or Master.

When the Submaster does not find a match in the Master for
its HD or DM number, the Submaster's entry is inserted in the
proper sequence. Since there is no star available in the Master
list at that point for a comparison check, such merged, or
inserted, entries could possibly introduce errors. However, as
subsequent Submasters are added, it is likely that such stars will
show up in other catalogues to be matched against the Master and
thus permit a check.

For ease of computer processing and to gain experience, we
started the project with a small catalogue, namely the <u>Yale</u>
<u>Bright</u> <u>Star</u> <u>Catalogue</u> (YBS) as our basic Master and then pulled
in the Submasters. In each case, a catalogue reference code was
carried to indicate the catalogue from which any piece of data

was recorded. In addition, the computer was given a specific
hierarchy of preferred sources for each of the following:
position, magnitude, spectral class, and proper motion.

The catalogues included in the Goddard Cross Index at the
moment are: the Smithsonian Astrophysical Observatory catalogue,
the Henry Draper catalogue, the Boss General Catalogue, Jenkins
Trigonometric Parallax catalogue, the U.S. Naval Observatory
Photoelectric catalogue, the YBS, the Strömgren-Perry uvby
catalogue, the Wackerling Emission-Line Objects catalogue, the
Batten Spectroscopic Binaries catalogue, Jaschek's catalogue
of Spectral Classifications and the Wilson Radial-Velocity
Catalogue. Each of these catalogues has its own peculiarities.
The more familiar we were with the machine-readable version of
each catalogue, and the more it had been used previously, the
fewer the difficulties usually caused by integrating that Submaster into the system. This does not mean that there would
necessarily be less stars in the error file of such a catalogue,
since, as explained earlier, the errors could be due to catalogues
merged into the Master at an earlier step. Obviously one of the
spin offs of such an exercise is to uncover errors which might not
have been detected without such comparison checks.

Working with the error files requires one's best sleuthing
abilities, using all the information at hand including magnitude
and spectral type as checkpoints. One must keep detailed records
of where one has looked and what one did or did not find. Each
error star has its own case history. Once we are convinced as to
where the error is, we make a computer-edit run to correct the
errors and to enter the correct catalogue ID's at that point.

Only the most basic data needed to identify each star has
been retained in the Goddard Cross Index. These are position,
visual magnitude, photographic magnitude, spectral type, proper
motion, catalogue ID, HD or DM, and reference catalogue for each
of the data items. Because the ID's themselves are preserved for
each catalogue, one can retrieve the full catalogue entry for each
star.

No weighting system for the data was used, only a hierarchy
of preferred catalogues for selecting the data source, when there
was more than one choice. Error ranges and critical remarks are
not included in the Goddard Cross Index since the data which is
preserved here is primarily for identification purposes.

Regarding the preface and explanatory notes we urge the user
to examine this material for each catalogue prior to using a
machine-readable version of the catalogue. If the preface or
notes are provided to the National Space Science Data Center
(NSSDC) with the catalogue tape, the Data Center sends out copies
of this material when it distributes a copy of the tape. The
NSSDC has agreed to act as a distribution center for North America.

The generation of the Goddard Cross Index is still in progress. The principal difference between this Cross Index and the Catalogue of Stellar Identifications (CSI) prepared by the Strasbourg group is that we have retained all the catalogue ID's explicitly instead of retaining only a flag to indicate member- ship in a given catalogue, as Strasbourg has done for some of its entries. At present the CSI contains more catalogues and also more stars than does the Goddard Cross Index. Although the ultimate end products of the two endeavours have basic similar- ities, because the sequence in which the catalogues were merged differs between the two projects, there is a greater chance of uncovering errors than if the same sequence of combination had been followed. Furthermore, a computer comparison of the "final" versions of the two Cross Indices will be an excellent check on the identifications made by that point.

V.   CRITICAL EVALUATION OF DATA

All responsible scientists know that the data compiled in catalogues should be evaluated critically. In the case of meas- ured quantities, not only a specific quantitative description of the quantity measured is required, but also a detailed description of how the measurement was made and what care was taken to make the measurement repeatable. In addition, it is desirable to relate the new quantity to values published in older catalogues and to screen the whole body of data for systematic and random errors.

Manipulation of a large data base is greatly facilitated by the use of a computer. The fact that large bodies of data can nowadays be handled rapidly and with little stress on the astron- omer makes it mandatory to consider the critical evaluation of data at the time of merging new and old data bases. We already possess a critical evaluation of many of the old sets of data. My concern is that we keep the old evaluations and develop eval- uations of the new data.

To talk about systematic and random errors is to imply that there is some value which is, by definition, "correct" or "best". This leads to a philosophical discussion about the meaning of "best". In reference to what characteristic property is the quantity best? All critical evaluations of a body of data should address this question. For a measured quantity such as radial velocity, magnitude or proper motion, "best" is usually defined as the mean value of all available results. The deviation of an individual measure from the mean value is called an error. Some- times weights are applied in forming the mean or best value. If so, one must ascertain whether the weights result in a bias. Ostensibly the weights are assigned to remove a bias or systematic effect believed to exist in the original data owing to some properties of the raw data or the measuring system. I think what one is trying to do by applying weights is to ensure that the total group of deviations from the mean value will follow the normal error function. However, in many astronomical studies one does not have a large enough body of data to justify the assumption that the errors (deviations from best value) will follow the normal error curve. Therefore, it is essential that the methods of obtaining the data and the selection of astronomical objects be well documented and that this documentation travel with the data whether it be machine-readable or not, so all factors affecting the data are known.

Before I deal with a specific case of critical evaluation, I shall review some of the characteristic properties to be considered when evaluating various types of astronomical data. Specialists in each field will recognize that what I mention here is in no way exhaustive. These possible sources of error, and more, are frequently discussed in the introductions to the older catalogues. It is somewhat disturbing that at present, information in these areas is frequently suppressed. I would like to see some way of preserving this information in machine-readable catalogues.

Spectral type   - What is the quality of the spectra? This may be
                   determined from the width of the spectrum, the
                   emulsion used, the dispersion and the level of
                   exposure.
                 - What line ratios or other criteria were used for
                   type and what for luminosity class?
                 - What are the standard stars for spectral type?

Colour classes   - How are they defined and what is their relation
                   to spectral type?

Photometry   - How were the extinction corrections obtained?
                   Are there any guiding errors indicating that
                   perhaps all the light did not fall through the
                   diaphragm or that other errors dependent on
                   telescope attitude occurred?
                 - What were the standard stars used to tie into
                   other photometric series? Were all these stand-
                   ards constant in light?
                 - How are the bandpasses and the zero point of the
                   magnitude scale defined? Is the scale truly
                   logarithmic over its whole range?
                 - Was any correction applied for light of the night
                   sky, background or nearby stars or for emission
                   nebulosity?
                 - How did the sensitivity of the detector degrade
                   with time?
                 - What is the change in the effective wavelength of
                   each bandpass with spectral type?

Positions   - How are the stellar positions tied to the

fundamental system of right ascension and declination? What sort of measuring errors occur? Are
any of them systematic?

Proper motions - What is the time base between several series of
plates?
- What are the measuring procedures and how are the
new results tied to old work?
Radio sources/ - What are the error boxes in position?
X-ray sources - What are the uncertainties in intensity?
- How is the bandpass defined and what is its eff-
ective frequency?

VI. AN APPLICATION OF THE TECHNIQUES OF CRITICAL EVALUATION

One of the first tasks to be done when one obtains a spectro-
gram of a new spectral region for a star or other object is to
identify the spectral lines which are present either in absorption
or emission. One compiles a list of the wavelengths of all
apparent lines and compares this list with laboratory lists of the
lines known in atomic and ionic spectra from all elements suspected to be present in the star or other object. If there are a
sufficient number of coincidences with the key lines of any one
species, one concludes that that species is probably present.
However, the problem is not simple. On the one hand, all the
measured lines may not be real and the list of observed features
may be incomplete. There may be false indications of the presence
of lines owing to noise in the spectrum record and some key lines
may be missed owing to data drop outs. On the other hand, the
laboratory lists may be incomplete, although they generally list
all strong lines, and lines of the various species frequently fall
at nearly the same wavelengths, causing blends. One bases a
decision that a certain species is present not only on obtaining
coincidence with many lines of the spectrum, particularly with
the strongest lines, but also with a consideration of the multiplet structure shown and on the general level of ionization and
excitation apparent in the stellar spectrum. Masking of the lines
of one species by those due to another is a serious problem,
because the spectral resolution of the observed spectrum is rarely
better than 0.2$\overset{\circ}{A}$ with the result that the wings of an observed
feature are at least 0.3$\overset{\circ}{A}$ wide, while the density of possible
lines is frequently 3 per $\overset{\circ}{A}$ or more. When the spectrum of a selected species, such as Fe II, contains many lines, many coincidences should be found before that species is considered to be
present. If, however, the species is believed to have a low
abundance, such as that for boron, and its spectrum contains few
lines in the spectral region under study, one must consider
seriously whether the few observed coincidences may be due to
chance. I have no quantitative rules to give for handling
problems like these. The following example will, however, demonstrate the problem.

I feel that computer matching of lists of observed lines
with lists of lines known in the laboratory is too insensitive a
technique for establishing what is present in the spectrum of an
astronomical source. The spectroscopist must survey all possibilities for identifications and make a judgement based on the
total evidence of degree of coincidence and possible masking.
Wavelength coincidence means that the observed and laboratory
wavelengths agree to within an amount $\Delta\lambda$, with $\Delta\lambda$ usually in the
range $\pm$ 0.2 to $\pm$ 0.4Å.

The problem of identifying the lines in the ultraviolet
spectrum of a B star as recorded from space presents a good
example of the critical evaluation needed when interpreting ultraviolet spectra of stars.

The observations with OAO-3, Copernicus, are made as follows:
In the Princeton experiment a narrow slit, behind which is a
photomultiplier, steps through the spectrum travelling along the
Rowland circle of the grating. This yields a set of points,
nominally a constant $\Delta\lambda$ apart, which give the count per dwell time
of 13.76 seconds. These intensity readings are arranged by computer in order of increasing wavelengths, they are corrected statistically for background counts due to corpuscular radiation
encountered in orbit and they are corrected for the stray and
scattered light in the optical system. They are then plotted as
a function of wavelength, the wavelength being corrected for the
motion of the star and of the earth and the satellite.

Figure 1 shows typical raw data for the star η CMa, spectral
type B5 Ia. The observed points have been connected with
straight lines. Lines have been drawn to represent the level of
the continuous spectrum and to represent the smoothed level of
stray light and noise originating in the detectors because of
noise in the electronics and noise due to encounters in orbit
with electrons and protons. The statistical uncertainty in a
count of N is $N^{1/2}$. On this account alone one wonders whether
small changes like the inflection appearing at 1048.2Å are real.
These data have been taken with the U2 spectrometer which records
intensities with a slit of 0.2Å wide at points separated by about
0.2Å. It is clear that there are many absorption lines which
overlap each other in the ultraviolet spectrum of a B type supergiant. The spectrum of the B6 III star, ζ Dra, is similar although
the lines are sharper.

Figures 2, 3 and 4 show regions of the spectrum of η CMa and
ζ Dra normalized to an adopted continuum after the corrections for
stray light, scattered light and particle noise have been made.
There are many absorption lines and suggested identifications
are noted above the spectrum record. It is clear that the resonance lines of abundant ions are stronger in the supergiant than
in the giant star. One wonders how reliable are the series

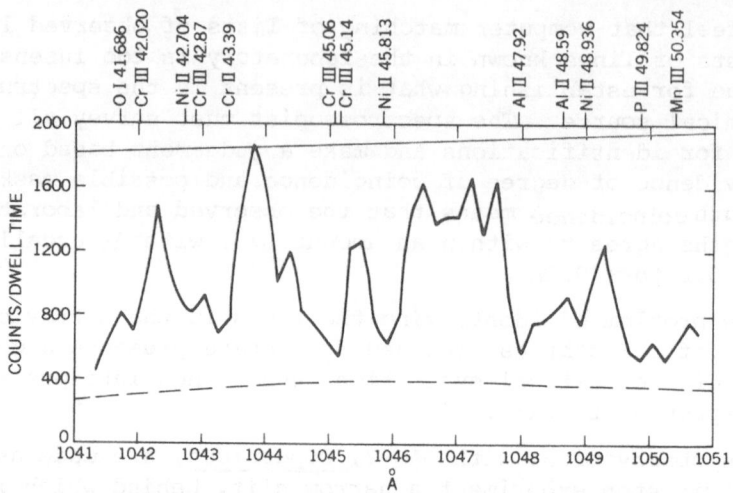

Fig.1- Raw data in the ultraviolet spectrum of η CMa, B5 Ia,
taken with the U2 spectrometer of the Copernicus satellite. The
observational points have been joined by a continuous line. The
upper thin line represents the continuum joining high points in
the spectrum while the broken line gives an estimate of the level
of counts due to background.

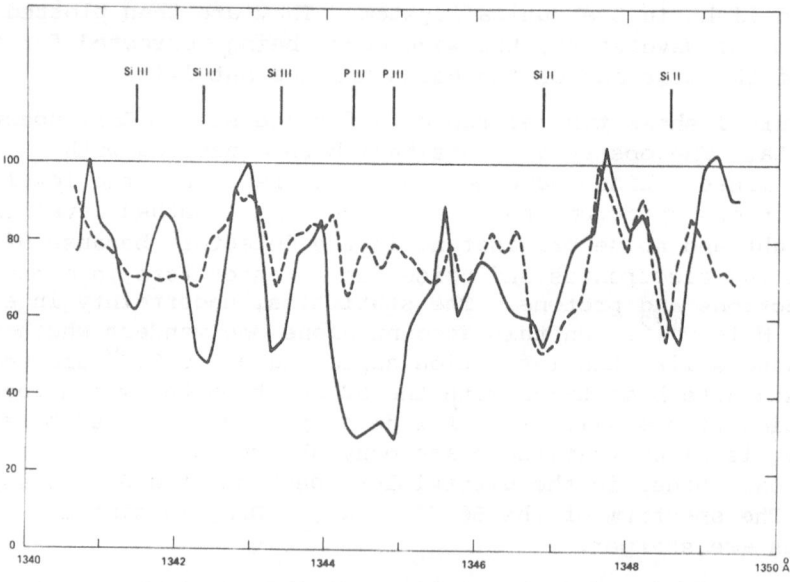

Fig. 2 - The normalized spectrum of η CMa (solid line) and of ζ Dra
(broken line) in the vicinity of the P III resonance lines derived
from tracings made with the U2 spectrometer of the Copernicus
satellite.

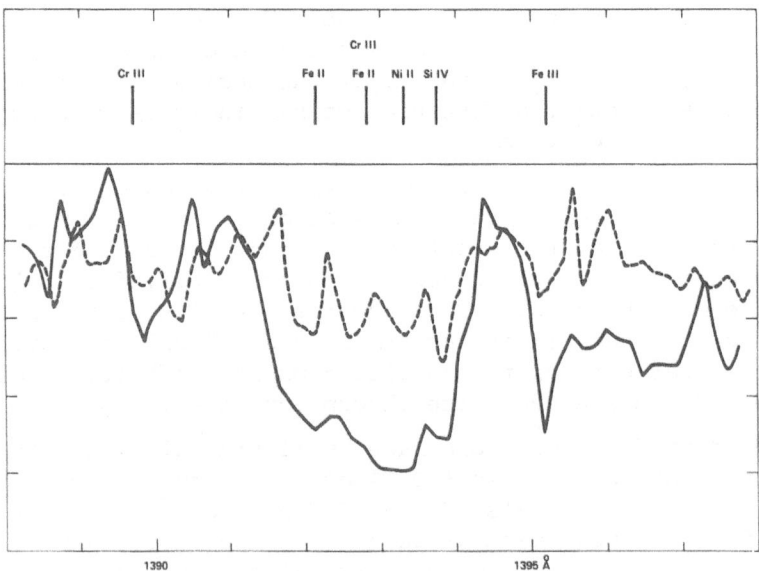

Fig. 3 - The normalized spectrum of η CMa (solid line) and of ζ Dra (broken line) in the vicinity of one Si IV resonance line, derived from tracings made with the U2 spectrometer of the <u>Copernicus</u> satellite.

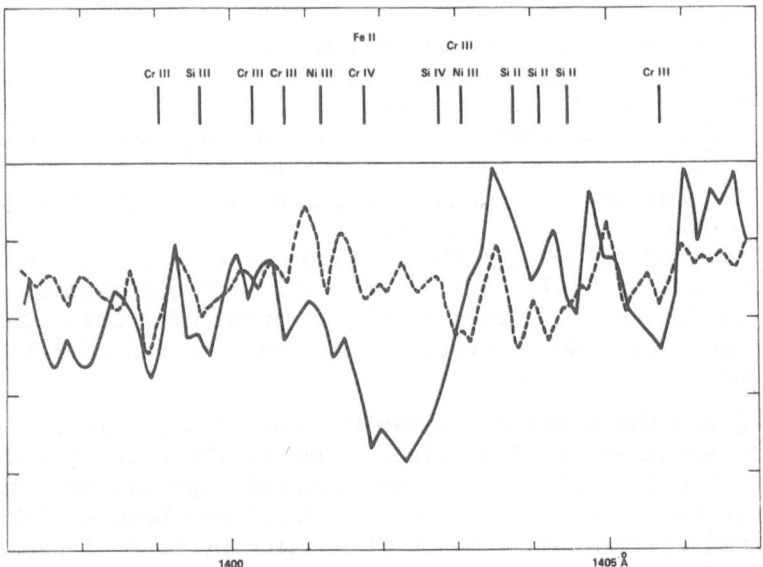

Fig. 4 - The normalized spectrum of η CMa (solid line) and of ζ Dra (broken line) in the vicinity of the other Si IV resonance line. The evidence for Si IV in ζ Dra is slight.

of identifications that have been made. Each dip and shoulder was
considered to indicate the probable presence of an absorption line.
After a survey of all the material it was decided that no emission
lines are present and that the high points indicate wavelengths
in the continuous spectrum.

The average observed line density in the range 1035 to 1425Å
is 1.7 lines per Å for ζ Dra. It is slightly less for η CMa.
However, over part of the spectrum the count is low and the statistical uncertainty in the number of counts received is 5-10 percent of the maximum count. This uncertainty plus occasional noise
spikes which are not eliminated by the statistical correction for
noise means that some of the smaller dips and inflection points
are false. Which are the false absorption lines?

The identifications were made by listing all lines from
likely ions that fell within ± 0.3Å of an observed line. Since the
third spectra of the metals (Ti, V, Cr, Mn, Fe) average 2.7 lines
per Å in the observed spectral range (see the wavelength lists of
Kelly and Palumbo 1973), and there are also many lines from the
second, third and fourth spectra of the light elements in this
range, most stellar lines have two or more possible identifications.
One must consider more data than the number of possible coincidences
in order to determine whether an atomic or ionic species is present.

Interesting spectra such as Be II and B II have a few lines
in the observed range. Can one decide that these ions are present
in ζ Dra? In the case of Be II, 8 lines are known between 1035
and 1425Å. All coincide within ± 0.2Å with stellar lines which are
attributed to other ions. One must conclude that Be II could be
present in ζ Dra, but that only a detailed spectrum-synthesis study
will prove whether it is there or not. In the case of B II, 9 lines
fall in the observed range, four of which coincide with lines attributed to other ions. In particular, the resonance line at
1362.461Å may contribute to the feature observed at 1362.40Å but
it is certain that Si III λ 1362.366 will also contribute, for
other Si III lines are strong in the spectrum of ζ Dra. Only spectrum synthesis will demonstrate if the contribution of B II is
significant.

In ζ Dra there are 690 lines between 1035 and 1425Å, six
percent of which are unidentified. Some of these are probably
spurious. The list of probable and possible species present comes
from an assessment of the amount of coincidence between laboratory
and observed spectra, the amount of masking by other lines and a
knowledge of what level of ionization and excitation occurs in the
atmosphere of the star.

Spectra from the International Ultraviolet Explorer (IUE)
satellite will be output as tracings which have been prepared by
computer from the digitally read charge on the target of the SEC
camera tube. This data will require careful evaluation to demonstrate what are true stellar or interstellar spectral features and

what are spurious features due to noise in the system or to peculiarities of the detectors. Since the spectral element of resolution is about 0.12Å in the high resolution mode for IUE, it will
not be possible to measure accurately the profiles of the very
sharp interstellar absorption lines found by Copernicus. In the
case of the low resolution spectra recorded by IUE, the spectral
element of resolution is about 6Å. Clearly in this case every
observed feature in the ultraviolet spectrum of a star will be a
blend of a number of contributing spectral lines. It will be
possible to determine the trends with varying spectral type of the
blended strong features in ultraviolet stellar spectra, but it will
not be possible to identify uniquely the individual lines contributing to each blend.

VII. SUMMARY

The critical evaluation of data is an activity that every
responsible scientist engages in. In the past, when data was published chiefly in books, the results of the critical evaluation
were usually presented in a preface. However, many old catalogues
are now out of print although the data contained in them continue
to have validity to some level. The problem of making old catalogues more widely available can be solved by transcribing the old
catalogues into a machine-readable format. However, this action
does not solve the problem of transferring with the catalogue
either the original critical evaluation or a new evaluation made
with reference to modern data. In addition, one must be alert to
sources of error or difficulties of interpretation due to computer
peculiarities.

New catalogues can be prepared efficiently by computer and
the use of a computer will facilitate the intercomparison of
various data sets to establish criteria which lead to an evaluation
of the scientific value of the data. Although computers can be
used to find and correct errors of format and of content, they
cannot provide critical evaluation of the data. That is an
activity which must be done by the scientist. Astronomers should
see to it that adequate critical evaluations of computer-processed
catalogues are published in the astronomical literature where they
are accessible to all who would use the machine-readable catalogues.

REFERENCES

Blanco, V. M., Demers, S., Douglass, G. G. and Fitzgerald, M. P.
    1968, Pub. U.S. Naval Observatory, 2nd Ser., XXI.

Kelly, R. L. and Palumbo, L. J. 1973, Atomic and Ionic Emission
    Lines Below 2000Å: Hydrogen through Krypton, NRL Report 7599,
    U.S. Government Printing Office, Washington, D. C.

Whipple, F. L. 1966, Smithsonian Astrophysical Observatory Star
    Catalogue, Parts 1-4, Smithsonian Publication 4652, U.S.
    Government Printing Office, Washington, D. C.

CRITICAL EVALUATION OF PHOTOMETRIC DATA

B. Nicolet and B. Hauck

Institut d'Astronomie de l'Université de Lausanne
and
Observatoire de Genève, Switzerland

ABSTRACT

Some problems encountered in the homogenization of photo-metric data (especially for the $uvby\beta$ and UBV systems) are presented: sources of inhomogeneity, method to obtain a homogeneous photometric value, establishment of a reference list, etc.

In many cases a homogeneous photometric value is necessary, but cannot be simply an arithmetical mean, and for this purpose it is necessary to make a critical evaluation of the data.

1. INTRODUCTION

The user of photometric (or other) data will not necessarily need a literal compilation of measurements in a given photometric system, but he will prefer often homogeneous data, in particular if he is not a photometrist.

In this paper, we discuss the methods and problems in preparing lists (more precisely files in magnetic tapes) of such homogeneous data.

Today, we will turn our attention to the UBV and $uvby\beta$ systems, but of course, we apply the method described below to all other photometric systems.

This work (compilation and homogenization of photometric data) is undertaken in the framework of the Stellar Data Centre (S.D.C.) (Hauck, 1974; Hauck and Jung, 1974).

All our catalogues on magnetic tapes are available from the Stellar Data Centre at Strasbourg (see Information Bulletin No. 8, p. 26).

C. Jaschek and G. A. Wilkins (eds.), Compilation, Critical Evaluation, and Distribution of Stellar Data. 121-124.
Copyright © 1977 by D. Reidel Publishing Company, Dordrecht-Holland. All Rights Reserved.

2. SOURCES OF INHOMOGENEITIES

When many observers make and reduce measurements in a photo-metric system such as UBV, they use various instruments and techniques.

This diversity introduces unavoidable inhomogeneities be-tween published results.

Among the sources of such inhomogeneities, the following can be mentioned:

Differences between:

- photomultiplier
- filters
- extra-atmospheric reductions
- use and choice of standard stars (perhaps lack
     of standards)
- optical properties
- cooling of photomultiplier

3. THE REFERENCE LIST

In order to detect the inhomogeneities and to estimate their importance, we must first prepare a reference list which will be homogeneous and will contain a variety of stars (i.e. of various spectral types, luminosities, apparent magnitudes, positions).

The necessity of homogeneity is obvious in such a list.

Concerning the variety of the objects contained in the ref-erence list, we can remark that the estimation of systematic or standard deviation between a given list of measurements and the reference list becomes significant only if the intersection of these lists is sufficiently substantial (say, more than 20 stars).

Thus, a reference list plays a more distinct role than a standard list (used for measurements) and it should be more ex-tensive.

For example, in the $uvby\beta$, Lindemann and Hauck (1973) have included in their reference list some measurements of Crawford.

In the case of UBV, the reference list of Nicolet (1976), in-cluding the measurements of Johnson and those made with similar instruments and techniques, contains more than 13,000 stars.

Of course, variable stars, which disturb the estimation of deviations, have been eliminated from the reference lists.

4. COMPARISON BETWEEN LISTS

Each list is compared with the reference list.   Generally the computer programme gives for the ith colour C^i (= magnitude or index of colour)

- the mean deviation of ΔC^i
- the standard deviation of σ^i
- a (multi) linear regression such as

$$C^i_{ref} = a^i_o + a^i_i \, C^i_{list} + \sum_{j \neq i} a^i_j \, C^j_{list}$$

where $\quad a^i_i \approx 1, \; a^i_o \approx o$ and $a^i_j \approx o$ if $j \neq i.$

5. HOMOGENIZATION

The numerical results of the comparison (i.e. ΔC^i, σ^i, a^i_o, a^i_i, a^i_j) can lead us either

- to eliminate the list, or
- to apply to it systematic corrections, or
- to conserve its measurements as published.

In both the second and third cases, the comparison of the list with the reference list permits us to attribute a weight w_j to the list.

For each star, the homogeneous i^{th} colour is obtained by the formula

$$\overline{C}_i = \frac{\sum\limits_j w_j \, n_j \, C^j_i}{\sum\limits_j w_j \, n_j}$$

with
  j rank of the list
  w_j weight of this list
  n_j number of measurements as indicated by the author
  C^j_i i^{th} colour in the list for the star with eventual systematic correction.

We can check the conformity of our list with the rest of the catalogue. This "rest of the catalogue" is surely less homogeneous than the reference list, but this check often reveals disagreement in remarks (variability, binarity) or in measurements. The critical determination of the causes of this disagreement (variability of the stars, inhomogeneities, errors) is important, but a tedious and delicate task.

6. FINAL CATALOGUES

For each photometric system we have, or are preparing, a catalogue on magnetic tape containing at least two parts:

- the first including the measurements as compiled from the literature,
- the second with the homogeneous data calculated with the method described before.

The final version will be sent to the S.D.C. at Strasbourg for distribution among the astronomical community.

In the case of the $uvby\beta$ system, this method has been used by Lindemann and Hauck (1973) with the lists of Crawford and Mander (1966), Crawford et al. (1970) and Crawford and Barnes (1971) as a reference list, as previously mentioned.

A second version has been published by Hauck and Mermilliod (1975). A third one is in preparation which will contain 7140 new measurements, including those of Grönbech and Olsen (1976).

For the UBV, the reference list (Nicolet, 1976) is ready on magnetic tape and we also have a tape containing all measurements in this system (more than 70,000) (Mermilliod and Nicolet, 1976), namely those included in the catalogue of Blanco et al. (1970) and those published later.

For some time we have been attacking, with the method described before, the tedious problem of homogenization of the UBV measurements included in the file.

REFERENCES

Blanco, V.M., Demers, S., Douglass, G.G. and Fitzgerald, P.M.: 1970, *Publ. U.S. Naval Obs.*, 2nd Series, 21.
Crawford, D.L. and Mander, J.M.: 1966, *Astron. J.* 71, 114.
Crawford, D.L., Barnes, J.V. and Golson, J.C.: 1970, *Astron. J.* 75, 624.
Crawford, D.L. and Barnes, J.V.: 1971, *Astron. J.* 75, 978.
Grönbech, B. and Olsen, E.H.: 1976, to appear in *Astron. Astrophys. Suppl.* 25, 2.
Hauck, B.: 1974, *Bulletin CDS*, Strasbourg, No. 6, 1.
Hauck, B. and Jung, J.: 1974, *Astron. Astrophys. Suppl.* 16, 289.
Hauck, B. and Mermilliod, M.: 1975, *Astron. Astrophys. Suppl.* 22, 285.
Lindemann, E. and Hauck, B.: 1973, *Astron. Astrophys. Suppl.* 11, 119.
Mermilliod, J.C. and Nicolet, B.: 1976, to appear.
Nicolet, B.: 1976, to appear in *Bulletin CDS*.

SOME THOUGHTS ON ASTRONOMICAL DATA FILES

William P. Bidelman

Warner and Swasey Observatory, Case Western Reserve
University, Cleveland, Ohio 44106, U.S.A.

Several astronomers were relaxing after a hard day at the
I.A.U. when the talk got around to their research activities.
One said, "There is a red star found in the Lowell proper-motion
survey that seems rather interesting - it is G232-75, which has a
proper motion of almost a second of arc per year. Since Giclas'
catalogue doesn't give any spectral type I'm planning to go to
Kitt Peak to observe it. The trip only costs about 300 dollars,
and besides it's fun to visit Arizona."

Second astronomer (a photometrist): "Don't bother; UBV
photometry is published for this object in Johnson and Morgan's
classic paper (Ap. J. 117, 313, 1953). This will tell you all
you need to know. I suppose maybe they give a spectral type for
it too but I usually don't pay any attention to those."

Third astronomer (a spectroscopist): "I take umbrage at
that remark. Actually, Giclas identifies the star as BD +56°2783.
With some effort you will find that it is not in the Henry Draper
Catalogue, but if you are clever you will find that it is in the
Henry Draper Extension, and assigned type M. However, Johnson
and Morgan call the object dM4 + dM6, thus implying that the star
is double. The types are indicated as being due to Kuiper, but
where he got them is beyond me. Also they refer to the star as
Kr 60, which I suppose means Kuiper 60, though I'm not up on all
of these oddball designations."

Fourth astronomer (a binary chap): "Really, now, everybody
knows that Kr 60 means Krüger 60, not Kuiper 60! Since the star
is obviously close to the sun, you should have looked it up in
Kuiper's famous (Ap. J. 95, 201, 1942) paper on the nearest stars.

C. Jaschek and G. A. Wilkins (eds.), Compilation, Critical Evaluation, and Distribution of Stellar Data. 125-129.

Joy and Abt (Ap. J. Suppl. 28, 1, 1974) give newer types of dM3.5 and dM4.5e, which may or may not be better. The binary has an orbital period of about 44 years. The only thing I don't know is who this joker named Krüger was. I looked him up in the American Astronomical Society membership list and he wasn't there. Maybe he didn't pay his dues."

Fifth astronomer (an older man): "Nonsense! If you were more scholarly you would know that Adalbert Krueger was the director of the Kiel Observatory from 1880 till 1895 and also editor of 40 volumes of the Astronomische Nachrichten. In his younger days he worked on the Helsingfors-Gotha +55° - +65° zone of the Astronomische Gesellschaft catalogue and in the footnotes to that work he listed some new double stars that were found with the meridian circle. But strictly speaking he doesn't deserve to have his name attached to the stars that you are talking about; the pair that he discovered was a rapidly-separating optical pair of no interest whatever. What really happened was that Burnham, at Lick, decided to look at all of Krueger's new doubles, and in doing so he found that the primary of Kr 60 was itself a 2" double. The "60" designation comes from the fact that this star was the sixtieth object in the list of new doubles that Burnham made up from the A.G. catalogue. Krueger never published such a list at all!"

Sixth astronomer (obviously very bored): "You guys are all missing the point. The only thing of real importance about this binary is the fact that the fainter component is a flare star called DO Cephei."

Seventh astronomer (who has just awakened from his slumbers at the mention of the words "flare star"): "Say, fellows, if there's any chance that your star is an X-ray source we'll fly a special satellite just for it. Uh, by the way, would you mind telling me where it is? You know I've got a terrible sense of direction."

(the end)

It is clear from my tale that one cannot get the full story on any astronomical object merely from its presence in any specialized catalogue; one needs also bibliographical information on the sources of data and on the uses that have been made of it.

My main concern today, however, is not the desirability of a general bibliographic astronomical data file, or even the details concerning its establishment or use - it is the necessity of making the data contained in the file as accurate as possible. This means, of course, as accurate as humanly possible, and I want to stress the fact that the intervention of human intelligence is essential if we are to propagate truth, not error. Now of course

we all know that published astronomical data are extremely
inhomogeneous in quality, to put it mildly, and appropriate
account of this fact should be taken by the conscientious data
compiler. But here I am not so much thinking of this as of the
numerous typographical or other more significant errors, such as
erroneous or dubious nomenclature, magnitudes, spectral types, coordinates, etc. that all too often occur in the literature. All
of us who compile astronomical data can recite hair-raising
stories of bloopers that have long gone undetected. We owe it
to our successors not to perpetuate this sort of thing.

I suppose that the ultimate blame for typographical errors
can be laid on the shoulders of the inventor of printing, who I
understand spent some time in this city. The best thing about a
lead pencil is the eraser at the other end, and the worst thing
about printing is that corrections have to be made later. And
today, when we are likely to introduce further errors into data
compilations through card punching and other computer operations,
we are in the unfortunate position of being able to propagate
errors faster than ever before.

Anyone who has seen any amount of wordage through the press
has experienced the most annoying kind of typographical errors -
those for which he himself is not responsible. Editorial offices
can do strange things over which we have little or no control. I
recall a few amusing instances of this sort of thing. As you
probably know, the headings at the tops of the right-hand pages
or articles are usually supplied by the editors. On one occasion
Sahade and Struve (Ap. J. 126, 87, '57) were surprised to learn
that they had written an article on W. (sic) Serpentis rather
than on the variable star W Serpentis. Presumably the W. was
an abbreviation for the star's first name. And Anders Reiz
(Ap. J. 120, 342, 1954) was amused to see that his article
involved "Stars with Negligible Content," rather than, as intended,
"Stars with Negligible Content of Heavy Elements." George Herbig
was a similar victim when he was quoted in IAU Vol. 8, p. 807,
1952 as saying, in the case of the variable star BE Cas, that "a
companion about 25 inches distant in p.a. 260° is of spectral
type G." And finally, a delightful blooper implying a non-
existent author is to be found in the recent Monthly Notices 169,
7p, 1974 in which an article on stellar polarimetry is attributed
to V. George and S. J. Coyne rather than to George V. Coyne, S.J.
This got as far as A. and A. abstracts!

This sort of thing, of course, doesn't hurt anything, but unfortunately editors and printers also like to take liberties with
figures and figure legends. It is not unknown for diagrams to be
gratuitously interchanged. If one is extremely unlucky his photographs can be turned upside down or half of his article deleted.
Sometimes, of course, this can improve the paper.

Then there are the sorts of misprints that are apt to be one's own fault: I once (Ap. J. 113, 304, 1951) accidentally classified a G8 V star as B8 V; luckily no one read the paper, so my reputation did not suffer unduly. It is quite common to find transposed digits or interchanged lines or columns in tables, to say nothing of references that do not refer to what they are supposed to refer to. Some authors are more prone to this sort of thing than others, of course. Many such errors are easy to spot and can quickly be rectified by a knowledgeable reader, but in some cases it is not easy to decide whether the questionable datum is a misprint or is just plain wrong. Only the author knows for sure, and he may be dead or otherwise unavailable.

Some who peruse the literature with a critical eye may find the pursuit of faux pas a pleasurable diversion, but to others errors in published work are deadly serious. I have recently had occasion to regret that the Smithsonian Astronomical Observatory catalogue occasionally fails to add the little "a" that should follow some BD numbers (e.g. +59°2664). This is how you can end up with two different stars with apparently the same BD number. Also I recently found (IAU Var. Star Bull. No. 1138) that two innocent stars have been wrongly suspected of being variable for 70 years simply because of minor designation errors in an AAS meeting abstract. Unfortunately in the bulletin in which I reported this, the word abstract is misspelled!

Even such a fine work as the third edition of the Yale Catalogue of Bright Stars has a few errors--though very few for such a large job. Most of us are aware of the difficulties that computers have with such numbers as -0° 25', and thus know about the errors in sign of some of the galactic latitudes in this catalogue. But you may not be aware that the star HR 3104 is assigned a declination of +46°76', which is enough to give one pause, whereas its true declination is actually some 30° different. Also, do not bother to look up HR 365, as it is not there. You will find it in the 1940 edition. Another cautionary word: don't use the Bright Star Catalogue as a source for the apparent magnitude of Arcturus. The sign is wrong!

There are even stranger things in the literature. I have some recollection of tabulated right ascensions off by several hours of time, as well as considerable confusion between BD, AG, and Bergedorf numbers, X's and chis, between capital A's, small a's, and alphas, and zetas and xis.

There is a predictably great amount of trouble with stellar coordinates due to precession, which is certainly one of the astronomical phenomena that we could usefully do without. Coordinates are not infrequently half-heartedly precessed, or, even worse, attributed to erroneous or even unspecified equinoxes.

Some of the confusion in the literature can be cleared up only by recourse to new observations. My colleague Bruce Stephenson has recently published two general catalogues of carbon and S-type stars, in the Warner and Swasey Observatory publication series, and I paraphrase here his remarks concerning the S star EP Vulpeculae: EP Vul, then nameless, was first recognized as an S star by Rust. The star has often been referred to as CE Vul, but the identification charts for the two stars plus Case and Hamburg objective-prism plates make it clear that the only S star in the region is EP Vul. Schaifers, looking for EP Vul because it had no published spectral classification, evidently found it and misclassified it as a carbon star; and in my carbon star catalogue I rejected it as carbon and corrected it to an S, although I was not then aware that EP Vul was already published as an S star under the name of CE Vul.

Eternal vigilance must be exercised in dealing with the literature. Recently my assistant casually, and without question, put into my data file some new variables in Monoceros that were assigned declinations in the +40's. Only the fact that I knew that the unicorn did not extend to a declination of +48° prevented us from making a serious mistake. Now I hope that no one will think that I am being unduly critical or making light of the splendid work that was and is being done by yesterday's and today's cataloguers and data compilers. But my experience has taught me that indeed published data must be critically examined before being accepted at face value and put into automated form where it will be speedily and broadly disseminated to less critical users. And while there is much that machines can do, the final stages of critical examination must be done largely by the human brain rather than by automation.

A final word should be said about errata as such. One of my former colleagues at Yerkes used to complain, when especially exercised, that the Harvard Announcement Cards were the only astronomical publication that didn't run errata. He thought they should! Actually, our treatment of the erratum problem in general is not satisfactory. Some authors send them in for publication six months or a year after the appearance of their papers; some never do--like the doctors they prefer to bury their mistakes. But even published lists of errata rarely fulfill their purpose, and I believe that perhaps, along with general astronomical data center, or data centers, we need an astronomical error center as well, that would see to it that errors in published material are as well known as the original papers. In any case, whether this proposal will result in anything definite, we who are concerned with data handling should put accuracy as our highest priority; I would rather be accused of knowing too little about something than of knowing too much about it that isn't so.

WORK ON THE GENERAL CATALOGUE OF VARIABLE STARS

B.V. Kukarkin, P.N. Kholopov,

Sternberg Astronomical
Institute

N.N. Kireeva

Astronomical Council,
USSR Academy of
Sciences

The Variable Stars Bureau of the Astronomical Council of the USSR Academy of Sciences presents this paper in order to provide information on our work with the hope that it will be discussed and recommendations to it will be made. More rational co-operation of our Bureau and the Strasbourg Centre de Données Stellaires is highly desirable. It should be based on the exchange of information gained in these centres in order to make both centres free from duplicating work.

The main task of the Variable Stars Bureau (VSB) consists in collecting and <u>analysing critically</u> all information in the field of investigating variable stars and other non-stable objects of our Galaxy (including variables in globular clusters). It is worth noting that no similar Bureau exists for treating information on variable stars in external galaxies. Our Bureau is not planning to extend our work in this respect yet.

The final output of our work is the General Catalogue of Variable Stars (GCVS). During thirty years of our work three editions of the GCVS (1948, 1958, and 1969-1971), 14 Supplements to them (1949-1976), 2 Special Supplements (1963-1972), 2 Catalogues of Suspected Variable Stars (CSV)(1951 and 1965), 19 designation lists of new variable stars (1946-1975) were published. VSB takes an active part in the publication of the journal "Variable Stars"; the 50th anniversary of the journal will be in 1978.

Now we commence working out the principles for compiling the fourth edition of the GCVS. We are not going to separate the issues of the CSV from the GCVS any longer. All the data of these publications will be collected in one series of books. We expect

C. Jaschek and G. A. Wilkins (eds.), Compilation, Critical Evaluation, and Distribution of Stellar Data. 131-133.

the fourth edition of GCVS to consist of 7 volumes, the contents
being as follows:

Volume 1. Catalogue of suspected variable stars containing
           data on approximately 9000 objects.

Volume 2. Catalogue of designated variable stars in the
           constellations And-Cyg, containing data on approx-
           imately 10 000 objects.

Volume 3. Catalogue of designated variable stars in the
           constellations Del-Per; also about 10 000 objects.

Volume 4. Catalogue of designated variable stars in the con-
           stellations Phe-Vul; also about 10 000 objects.

Volume 5. Bibliography for the four preceding Volumes, list
           of all variables and suspected variables arranged
           in the order of right ascensions for the equinox
           1950.0 (adopted in the catalogue), list of stars
           arranged according to type of variability, data
           on optically non-stable objects of nonstellar
           nature, lists of pulsars, supernovae, variable
           X-ray sources (it is not excluded that this Volume
           will consist of two books).

Volume 6. List of all stars in the order of some other equi-
           nox (probably 1900.0), to facilitate identifica-
           tions.

Volume 7. Various nomenclature tables of variable stars, with
           different entries (in the order of the Catalogue
           itself, in the order of Durchmusterung numbers, BS
           and HD numbers, numbers of the Smithsonian Obser-
           vatory Catalogue, in the order of preliminary des-
           ignations, etc.) (it is not excluded that this
           Volume will be also published in two books).

The final version of the plan will be adopted after the discussion of our project at the meeting of the IAU Commission 27 in
Grenoble.

The majority of scientific workers and experts of our Bureau
deal with a thorough analysis of all published literature and of
private communications we receive. Then goes the selection and
evaluation of information of interest for us; the information
is put on cards and magnetic tapes.

Serious complications in the work of the VSB arise from the
laborious job of checking co-ordinates, identifying stars by using
charts, etc. Despite all these checks till now there happen rare
cases of giving different designations to the same object, of wrong
identifications, etc. Usually all responsibility for these facts
rests with the authors who do not take trouble of accurate determination of co-ordinates (even to the tenths of arcminute) or of
preparation of good finding charts.

All above-stated makes it clear that our tasks are quite specific, and are different from those of the Strasbourg Centre and its Potsdam branch or the tasks of Sonneberg observatory connected with compiling the "Geschichte und Literatur".

Harmonious combination of efforts of these three centres could be very useful for all participants. For example, the existence of the General Catalogue of Stellar Identifications and of annual Supplements to it compiled at the Strasbourg Centre would help us to check our identifications and free us of this work in future. It will be useful for the Strasbourg Centre to get more operative information on new variable stars data. These data are being continuously changed and supplemented, but they are published only from time to time (time interval between the editions of the GCVS is on average 11 years, and that between the Supplements to the GCVS is about 3 years).

The work of Moscow VSB has independent significance and it does not duplicate the work of either the Strasbourg Centre or Sonneberg observatory. The results of our work are unique, and we are ready to discuss any forms of co-operation and mutual assistance.

We are sure that the main aim of both the Strasbourg Centre and our VSB is to find such ways of collecting information and of organisation of information exchange that will enable interested astronomers throughout the world to get with shortest time delays the most reliable information (with account of the latest publications or private communications) on objects of interest for them.

A CATALOGUE OF PHOTOMETRIC SEQUENCES
SUPPLEMENT NO.2

A.N. ARGUE   and   E.W. MILLER

Institute of Astronomy   Physics Department
University of Cambridge   Ricks College
                                 Idaho

Photometric observers have from time to time expressed the need for a convenient way of finding in the literature the photometric sequences they require.   This need is particularly felt in photographic photometry, and the increasing use of the Palomar Sky Survey in recent years, and more recently the ESO and SRC surveys, have given a new impetus to the subject.   In addition, the use of image tubes has led to a further demand for suitable calibration sequences.

At the Brighton IAU Meeting in 1970 Argue, strongly seconded by Bok, proposed to the Photometry Commission that a small working group be set up to produce lists of sequences.   The Commission President, Dr. Cousins, requested Argue and Bok to contact interested parties and compile a list.

A circular letter was sent out to astronomers known from a search of the literature and of the reports of observatories to be active in the photometric field, and from the replies received a catalogue was produced early in 1973, and later that year a supplement.   For each sequence the catalogue gave the equatorial and galactic coordinates, the name given to the sequence by its author, the angular extent on the sky, the number of stars, their range in brightness, the photometric scales (UBV, Hβ etc.) and a literature reference or alternatively the author's address.

In this work Argue and Bok were assisted by E.W. Miller, an associate of Bok's at Steward Observatory.   Altogether some 480 sequences were given, together with data on about 2000 stars scattered over large areas of sky.   A reference was also given to an earlier compilation by Sharov and Jakimova in Sternberg Trudi <u>40</u> containing some 800 entries of which about 450 are

C. Jaschek and G. A. Wilkins (eds.), Compilation, Critical Evaluation, and Distribution of Stellar Data. 135-139.

suitable for calibration sequences.   Sharov and Jakimova's
catalogue covers the period 1953 to 1968.   Ours begins about
1968, so altogether there is good coverage over the last twenty
years.   We do not however claim that our part of the coverage
is complete.

The main catalogue and supplement were circulated by us
principally to contributors and observatories.   Our stock of
400 copies is virtually exhausted, but copies are available from
the Centre de Données Stellaires, Strasbourg, details of which
are in Information Bulletin No. 9.

At the Sydney Meeting of the IAU in 1973 the Photometry
Commission sponsored a further supplement.   Bok had retired
about this time and his part was taken over by Miller.   Again
a literature search was made and 500 circular letters sent out,
200 by Miller to the N and S Americas, and 300 by Argue to the
rest of the world.   It is interesting to note the response.
Miller received 9 replies with data on 129 sequences.   Argue
received 14 and 164 sequences.   We do not believe this adequately
covers the potentially useful sequence photometry that has been
carried out in the past three years.   There are probably still
many more sequences available.   I shall return to this point
later.

Some observers replied to the circular letter by writing a
letter of encouragement, usually combined with a request for a
copy of the supplement when it is ready, but not supplying any
data.   I suppose where physical support is lacking we must be
content with moral support.

In addition to the 300 sequences just mentioned, our new
supplement refers to 200 open clusters catalogued by Mermilliod,
5000 stars in $UBVB_1B_2V_1G$ by Golay and 25 sequences in galactic
HII regions by Moffat.[1]

Altogether, taken in conjunction with the earlier catalogue
and supplement, the user has some 1400 sequences for immediate
reference and in addition, data on about 10000 stars scattered
over the sky.   The various sources from which the sequences
have been derived are listed in Table I.

TABLE I.   Distribution of sequences

Main Catalogue and Supplement 1	480
Supplement 2	300
Sharov and Jakimova	450
Mermilliod	200
Total	1430

Fig. 1 shows the distribution of limiting brightness for the sequences listed in Table I.   Most are seen to terminate at about 15^m, the limit for a telescope of aperture 1m.

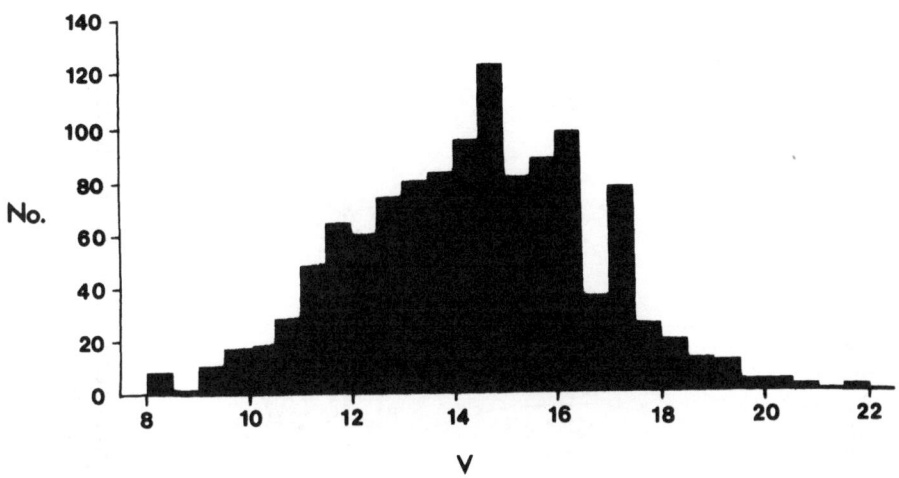

Fig.1.   Distribution in limiting magnitude of sequences in
          Table I.

The distribution over the sky is shown in Fig.2 which is an equal area plot in galactic coordinates.   Open circles denote sequences in which all stars are brighter than 14^m, and filled circles the remainder.   Apart from dividing our sequences into two approximately equal groups, 14^m is a convenient upper limit for magnitude estimation from the Palomar prints.   Below 14^m the image diameter varies linearly with magnitude to a close approximation down to about 19^m on the O prints.   This gives a useful way of determining brightness within this range because the slope does not show much change from area to area or chart to chart. The most important changes are in zero point, so that once this is fixed, say by a few stars of magnitude 14, brightnesses can be determined to an accuracy of $0^m.3$ of $0^m.4$ in the same area. This is very useful for QSOs.   The closed circles in Fig. 2 indicate in which regions useful sequences are to be found.

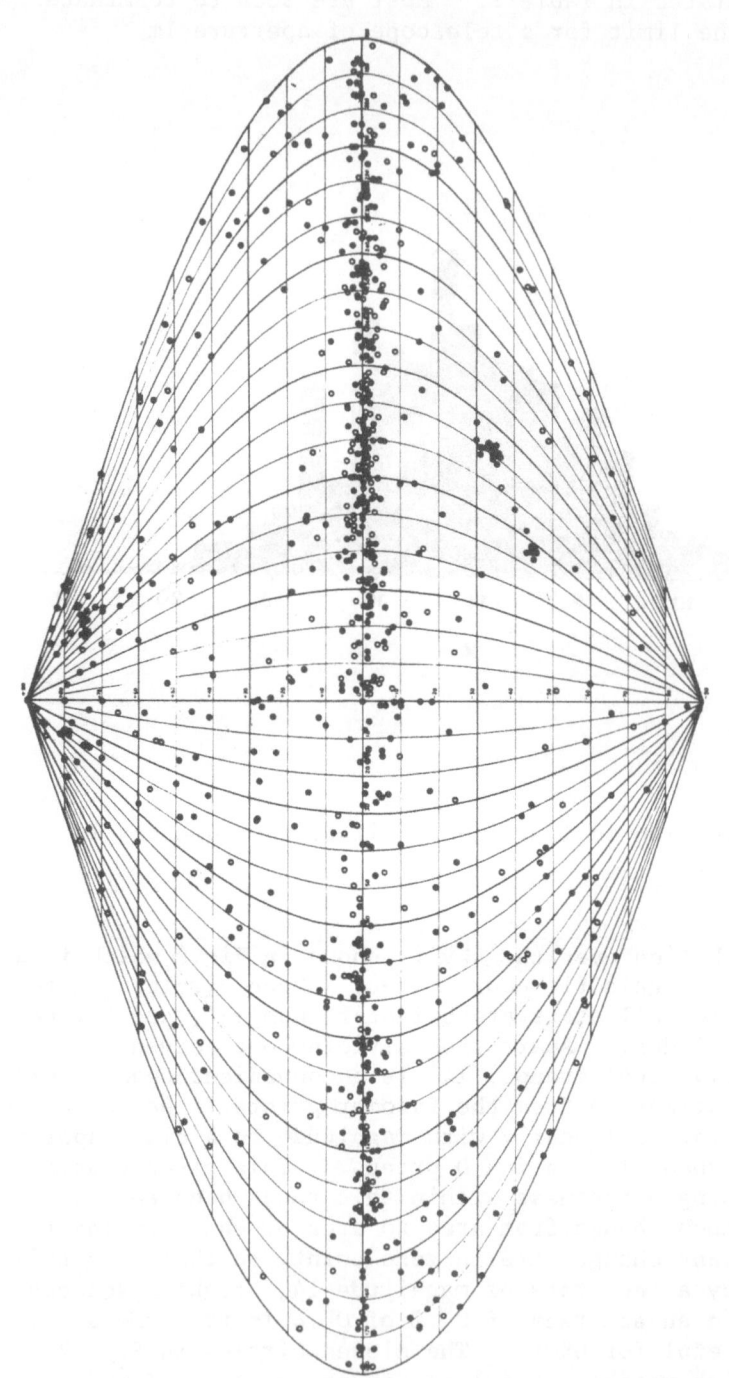

FIG. 2.   Distribution of sequences in Table I.   Open circles: sequences having all members brighter than 14^{m}.   Closed circles: the remainder.

The lack of completeness of our compilation raises the question of whether we have been compiling in the best way, or would we have done better to have compiled directly from the literature? Compiling from the literature can be very time consuming. Once a paper has been seen to contain useful data, it is easy to send the author a circular letter, but to extract the photometric information from the paper can take much longer: the right ascension and declination alone may require a search through several cross references. On the whole we prefer the more personal approach. It gives the author the chance to revise published material and to communicate in advance of publication. Much of it may of course be published eventually, so a literature search will achieve the same result in the end, but the current trend is against this. Journals are becoming increasingly reluctant to accept tabular data, and page charges are ever increasing. The convenience of having all your data ready at hand on your library shelves is rapidly passing away. Our catalogue goes some way beyond this, in establishing a direct communication between observers, but with steadily rising production and distribution costs this too is becoming difficult to continue, at least on the present privately financed basis. The answer lies in central data banks, clearly in this case Strasbourg. These are matters I am hoping the Photometry Commission of the IAU will consider, and I would very much appreciate the comments of this colloquium.

CATALOGUES OF SPECTROSCOPIC BINARIES

Alan H. Batten

Dominion Astrophysical Observatory, Herzberg
Institute of Astrophysics.

There are two centres at which papers on spectroscopic
binaries are filed and catalogues produced -- Toulouse, France
and Victoria, B.C., Canada. The aims of these two are somewhat
different. At Toulouse a running file of all references to
spectroscopic binaries is kept, and many workers have found the
comprehensive series of Catalogues Complémentaires (e.g.
Pedoussaut and Carquillat 1973) very useful. At Victoria, on the
other hand, we concentrate more specifically on spectroscopic
determinations of orbital elements, and our catalogue of them was
a direct successor of the five such catalogues published by the
Lick Observatory. While the Toulouse catalogues are comprehensive,
we at Victoria try to make a critical assessment of each orbit we
include, and we regard this as one of the most important aspects
of our catalogues. Such a critical catalogue obviously cannot be
published as frequently as are the Catalogues Complémentaires,
which, just as their name implies, are complementary to the
Victoria work. There is no fixed interval between the publication
of successive Lick-Victoria catalogues. The five from Lick
appeared at a mean interval of about a decade, but the transfer
to Victoria delayed the appearance of the Sixth Catalogue (Batten
1967) to almost twenty years after that of the Fifth (Moore and
Neubauer 1948). After another decade, the Sixth Catalogue is out
of date and out of print, and we are well advanced on the
preparation of a Seventh, which we hope will appear in 1977. The
Sixth Catalogue contained orbital elements for 737 spectroscopic
binary systems. About 200 (or 25 per cent) new systems have now
to be added, and about one per cent of those already in should be
deleted. In addition, new orbital information is available for
over 100 of the systems already in the Sixth Catalogue. As we
plan to keep the files open until at least the end of 1976, these

C. Jaschek and G. A. Wilkins (eds.), Compilation, Critical Evaluation, and Distribution of Stellar Data. 141-145.
Copyright © 1977 by D. Reidel Publishing Company, Dordrecht-Holland. All Rights Reserved.

figures may well be exceeded when the <u>Seventh Catalogue</u> appears
in print. The past decade has seen great activity in the field
of spectroscopic binaries -- including the valuable work of
Popper on accurate mass determinations from eclipsing binaries
that show two spectra -- but nevertheless, nearly ninety years
after the discovery of the first spectroscopic binary, still
fewer than a thousand of these objects have known orbital
elements.

The Victoria files have been only on handwritten cards,
until now, and updating them has been slow. Mr. J.M. Fletcher
is working on a punched-card system that will make updating the
files fairly routine, and speed up the production of the new
catalogue. This will be especially useful when the time comes
to hand over the task of preparing these catalogues to someone
else, or even to another institute. Mr. Patrick Mann is undertaking the laborious task of punching cards for all the existing
data. The new catalogue will have much the same format as the
existing one has, but we are planning to round off the values of
the various orbital elements to specified numbers of significant
figures. We regret rounding off rather than printing the values
the original authors gave, but many observers quote more
significant figures than are justified. On the other hand,
rounding off will enable us to introduce an extra column on one
page and to reinstate the values of <u>a</u> sin <u>i</u>, which were not
included in the <u>Sixth Catalogue</u>. We are trying to record as many
magnitudes as possible on the photoelectric V system, and spectral
types on the MK system, and this will make necessary some minor
modifications in the presentation of these quantities. In this
part of our work we shall be as much users of data compilations
as producers.

We expect to use three cards for each set of orbital
elements printed in the <u>Catalogue</u>. The first card identifies
the star by H.D. number or other designation, coordinates for
1900, photoelectric V (or best available) magnitude -- together
with the range of variation for eclipsing stars, and MK (or
best available) spectral type. The second card contains the
elements -- period in days, Julian Date of periastron (or other
suitable epoch), longitude of periastron in degrees, eccentricity,
semi-amplitude of velocity variation, systemic velocity (both in
km s^{-1}), and an estimate of the quality of the orbit. The third
card contains the bibliographical reference. Second and
subsequent sets of orbital elements can be punched on two
further cards each, but only the selected set of elements will
be printed. Precession, the mass-function (or minimum masses),
and the major semi-axis will be computed when a printed copy is
made. We have already uncovered several errors in the <u>Sixth
Catalogue</u> that arose from incorrect computation of these
quantities (usually by the original author). On the other hand,
we shall now have to make extra checks for typographical errors
in the published values of the semi-amplitude.

We intend to continue with a five-point scale of quality ratings of the kind used in the Sixth Catalogue and applied in accordance with the criteria set out there, Ideally a quality rating should be objective in the sense that any two workers in the field would assign the same quality to a given orbit. In practice, objectivity of that kind is hard to attain and there will always be some disagreement about border-line cases. One might resolve this by delegating the assessment of quality to a committee of experts, but the judgements of a committee seldom inspire confidence, even (or, perhaps, especially) if it is composed of experts. The best way is probably to let one person, or a close group of colleagues do their best. The subjective opinions of those who know their job are not without objective value. One can try to quantify the various criteria that are used in reaching a quality assessment and weight the score for each criterion according to an agreed scheme. This is not always easy, however. A quantitative measure of the distribution of points along a velocity curve, for instance, is hard to devise; and the relative weights to be attached to this criterion and the dispersion of the spectrograph, say, are themselves a matter of subjective opinion. I still think the best way is to rely on one's "feel" for the quality of an orbit determination, and this cannot be quantified. Fortunately most sets of orbital elements classify themselves. Provisional or preliminary solutions are automatically classed d or e, as are many determinations made as a result of survey programmes in which full attention cannot be given to the achievement of the optimum distribution of observations. There are also some systems (e.g. U Cephei) in which the intrinsic uncertainties are so great that no matter how good or numerous the observations may be, only poor-quality orbital elements can be obtained. On the other hand, systems observed several times are bound to be classed as a or b unless independent determinations disagree strongly. Average determinations of orbital elements, which is what most of the rest are, belong in class c.

Quantities like orbital elements are not observed directly, but deduced from observations. Sometimes later investigators will reduce the data again in their own way, and present the maker of catalogues with two somewhat different sets of orbital elements based on exactly the same observations. It is difficult to know which set to choose for the catalogue. Our aim is always to list the best determination of orbital elements. Obviously those who recalculate the elements usually believe they have effected an improvement, otherwise they would not have undertaken the labour of calculation. It is not, however, always clear that they are right. Lucy and Sweeney (1971) recomputed orbital elements for many of the binaries in the Sixth Catalogue. They were interested in the reality of small eccentricities, and argued that many binaries listed as having slightly elliptical orbits had, instead, orbits that on the

presently available evidence could not be distinguished from
circular ones. They devised a criterion for testing the reality
of a calculated eccentricity and, using it, rejected those
obtained for many binaries. They derived new orbital elements
for these systems, assuming the orbits to be circular, and
maintain that these new values should be adopted. Their method
of computation is uniform (although the observations themselves
are heterogeneous) and it is tempting to make wholesale use of
their results. There are, however, both ethical and scientific
reasons for hesitating to do so. The ethical one was summed up
by Moore and Neubauer (1948) who said that the original
investigator "is entitled to the credit because of his greater
contribution". Although credit is supposed to be of secondary
importance to a scientist, his standing is often measured, these
days, by the number of papers he publishes, or even the number of
times they are cited. The compiler of a catalogue ought not to
ignore that fact. The scientific reason is the question whether
or not the elements obtained by Lucy and Sweeney are always
improvements over the original values. While those investigators
are certainly right to maintain that many orbits of small
eccentricity are indistinguishable from circular ones, it does
not necessarily follow that the eccentricities deduced should be
ignored. If the observational errors obey the normal law, the
statement that e = 0.04 with a standard deviation of 0.02 has
the precise meaning that e is twice as likely to lie between 0.02
and 0.06 as to lie outside that range. This is different from
saying that the orbit is circular. It is true, as Lucy and
Sweeney point out, that the light curves of many eclipsing
binaries show the orbits to be circular although the velocity
curves correspond to elliptical orbits. These eccentricities
are spurious in the sense that they do not correspond to the
geometrical reality of the orbit, but they may nonetheless
measure a real distortion of the velocity curve. Spurious
eccentricities can arise from accidental errors of observation
or from systematic ones. It is often difficult to know which
kind of error is acting in a specific case, and we hesitate to
suppress the information that may be conveyed by a small
eccentricity, even though we recognize that, naively interpreted,
that information is often misleading. We have adopted the
results of Lucy and Sweeney when they seemed to us to be real
improvements over the original result, but otherwise have
relegated them to the notes at the back of the Catalogue. The
problem is discussed at length here because it is, perhaps,
typical of problems that arise in this sort of work and merits
some general consideration.

    We shall continue to provide notes on virtually every system
in the catalogue as such notes seem to us an essential part of
any critical catalogue of the kind we have been discussing.
Since detailed notes lend themselves less readily to presentation
on punched cards than do the numerical data, the notes perhaps

form the principal justification for issuing the catalogue in book form. The cost of the Sixth Catalogue (reckoned only from the printing costs) was about $10 per copy: that of the Seventh will undoubtedly be appreciably higher. We have to balance these costs against the convenience of a catalogue in printed book form. For the Seventh Catalogue the balance will probably be in favour of a publication in essentially the same format as that used for the Sixth Catalogue. If present trends continue, however, it may be necessary to look for cheaper ways of producing further catalogues in the series; perhaps the solution is to provide supplements on tape or punched cards.

References

Batten, A.H. 1967 Publ. Dom. Astrophys. Obs. 13, 119.
Lucy, L.B. and Sweeney, M.A. 1971, Astr. J. 76, 544.
Moore, J.H. and Neubauer, F.J. 1948, Lick Obs. Bull. 20, 1.
Pedoussaut, A. and Carquillat, J.-M. 1973, Astron. Astrophys.
    Suppl. 10, 105.

THE PRINCETON-PENNSYLVANIA-FLORIDA CARD CATALOGUE OF ECLIPSING VARIABLES

Frank Bradshaw Wood

Rosemary Hill Observatroy
University of Florida
Gainesville, Florida 32611, U.S.A.

For many years, the study of close double stars seemed some-what remote from many astronomical problems. The specialists in the field pursued it with enthusiasm, regarding each light or velocity curve as a particular challenge and each system as unusual and interesting in its own right. However, the relation to most other branches of astrophysics sometimes seemed remote. In favorable cases, masses and radii (and hence mean densities) could be obtained and the mass-luminosity relation could be strengthened; even this advantage was partially negated when it was early realized that in many cases the fainter components were notorious violators of the relation established by visual binaries (i.e., systems well separated compared to those discussed here). Various complexities, connected chiefly with interaction effects between the close components, prevented their use for the extremely precise determinations of limb darkening which at one time seemed possible. Other difficulties have prevented the determination, with the precision once comtemplated, of details of internal structure from the apsidal rotation.

However, in recent years, close double stars have become important in various other branches of astrophysics. At least some x-ray sources have been identified as members of close binary systems. Some are intermittent emitters of radio waves. Most eruptive variables seem to be components of close double stars. The evolution of a star when it is significantly affected by the presence of another is strikingly different at many phases from that of a single star. The net effect of these and other developments has been brought into the field many astronomers who have not previously studied close binaries; the purpose of this communication is to inform them of the existence and nature of this catalogue.

C. Jaschek and G. A. Wilkins (eds.), Compilation, Critical Evaluation, and Distribution of Stellar Data. 147-150.

The card catalogue was started more than fifty years ago at
Princeton University by Raymond S. Dugan. It was continued by
Newton L. Pierce and then at the University of Pennsylvania and
since 1968 at the University of Florida by myself, with the aid
of a series of competent graduate students and an occasional post-
doctoral fellow.

The stars are listed not only in the conventional variable
star designation (e.g., R Canis Majoris, AR Lacertae, etc.) but
also whenever appropriate by discovery number (e.g., 77.1929,
255.1930), HV (Harvard Variable), HD (Henry Draper Catalogue
number), BV (Bamberg Variable), SVS (Soviet Variable Star), VV
(Vatican Variable), OV (Oklahoma Variable), S or So (Sonneberg
Variable), and various other designations.

The information listed is of the conventional type: author,
journal, volume, page, and year. For a full length paper, either
the title or a brief summary of the article is given. New observational data such as depths or widths of minima are of course
recorded, as well as new light elements and the results of solutions
of light or velocity curves. Individual times of minima are frequently recorded, especially if the publication is one not found
in all astronomical libraries.

For some systems, a considerable volume of publication exists.
In the case of U Cep, for example, 35 cards (20 cm x 12.5 cm each)
have been filled with references and data; to illustrate the modern
volume of work, even going back four cards only takes in the work
of the past two years. HZ Her, discovered only in 1972, has more
than 18 cards of closely written data references and card 19 is
now almost filled.

In addition to the published material, a good deal of unpublished material is included. This ranges from completed typescripts
sent at the time the paper is submitted to a journal to simple
statements that the system has been observed. (Because of experience, we seldom include a statement that someone "plans to observe";
normally some observations must actually have been made before a
statement is included). Further, abstracts of papers at meetings
are noted, and observatory reports are treated in much the same way
as letters; we record work actually done or going on, but not
expectations or hopes. As a check on the catalogue, we make regular
comparisons with the "Bibliography and Program Notes" of Commission
42 prepared under the direction of Gunnar Larsson-Leander and with
supplements to the General Catalogue of Variable Stars.

DISTRIBUTION OF DATA

The distribution of information is carried out in three different ways. First, material is distributed in reply to personal requests merely by xeroxing the cards and sending the copies. We emphasize that we make every effort to keep the catalogue complete; it is not intended to be critical. Those using it are presumably professional astronomers, or students working under their guidance; they can read critically the publications and form their own judgments. (It may be well to emphasize the danger of accepting any value without checking the source; see E. W. Weiss, Observatory 96, 9, 1976, on one error perpetuated for years because it was carelessly accepted in a published catalogue.) Our purpose is to give as complete a set of references as possible. Answer to requests have usually gone out no later than the day after the request was received.

A second method of distribution of information has been the publication at irregular intervals of "A Finding List for Observers of Eclipsing Variables". This was initiated by Dugan "as an aid to observers of eclipsing variables in selecting a program rapidly, easily, and without exasperation". Four editions have been published to date. The last by R. H. Koch, S. Sobieski, and F. B. Wood appeared in 1963 (Vol. IX of the Astronomical Series of the University of Pennsylvania), and another is needed. While some changes have been made in the format, we have adhered to Dugan's original plan of giving numerical data on the left hand page and historical references on the right. In addition to magnitudes, spectra, depths and durations of eclipses, and other data, we have listed for each system the most reliable ephemeris available at the time. Most of the columns are self-explanatory; those headed S and P simply indicate whether in the opinion of the authors, further observational work in spectroscopy or photometry is needed (*), or badly needed (!); the numerical superscripts are literature references and are identified on the last few pages. Again, the purpose of this is to aid observers to select rapidly systems of interest to them; it is expected that they will consult the original literature before commencing observation.

Finally, the material has served as the basis for related publications - i.e., publications that could not have been produced at all, or which would have taken very much longer to produce, had it not been for this data source. These include "A Catalogue of Graded Photometric Studies of Close Binaries", by R. H. Koch, M. Plavec, and F. B. Wood, "An Atlas of Light Curves of Eclipsing Binaries" by M. G. Fracastoro, and portions of reports of I.A.U. Commission 42 in 1952, 1964, 1967, 1970, and 1976.

FUTURE DEVELOPMENTS

The continuation of the catalogue at least in its present
form is essential. The volume of publication is such that inter-
ruption for even a year or two would make its resumption an
exceedingly expensive and time consuming task. For example, in
the listing of articles of general interest not connected with
one particular star - theoretical developments, new methods of
solution or models, general summaries, discussions of evolution,
etc. - one page (25 x 20 cm.) covered all the material published
in 1931 and 1932 (authors, journal references, letters, and occasional abstracts); later, about two pages were required for the
general publications of 1951 and 1952; for 1975 and the first
four months of 1976, ten full pages were required to list author,
journal reference, and title. A similar expansion has been found
for many individual systems and especially for those which are
x-ray sources.

The time has now come (and indeed is a bit overdue) when
the catalogue should be put on punched cards suitable for handling
by modern computers. Most of the data is ideally suited for this
treatment and even in the "remarks" it should not be difficult to
provide coded treatment. (Indeed, I am informed that at a number
of observatories, the most recent edition of the Finding List has
been so treated.) Ideally, this transformation should coincide
with the preparation of a fifth edition of the Finding List.
With cards so prepared, the preparation of annual or bi-annual
editions should be relatively simple. Shortage of sufficient
funding has been the chief reason for not having made the change
to date. Meanwhile, we intend to continue the catalogue in its
original form and will continue to supply information as requested.

(July 1976)

PLANS FOR A NEW EDITION OF THE BRIGHT STAR CATALOGUE

Dorrit Hoffleit   Carlos Jaschek

Yale University   and   Observatoire de Strasbourg

Almost any star catalogue becomes obsolescent as soon as it comes off the press. It is now twelve years since the third edition of the Yale Bright Star Catalogue was published in 1964. In the intervening years numerous important new results have appeared, notably the Jaschek et al Catalogue of MK Spectral Classes, the Lick Double Star Catalogue, the U.S. Naval Observatory Catalogue of UBV Magnitudes, the Abt and Biggs Bibliography of Radial Velocities, the Third Edition and three Supplements of the General Catalogue of Variable Stars, the SAO Catalogue with proper motions reduced to the FK4 system, several new lists of spectroscopic binaries, and a great many shorter compilations. Consequently well in excess of 20,000 entries in the third edition of the Bright Star Catalogue need up-dating.

The Centre de Données Stellaires and Yale are planning to collaborate on a new edition, incorporating major revisions. Some cross-reference columns may be deleted as being less useful for identification purposes than in the past (e.g., ADS numbers), and other columns may be added if sufficient new data so warrant. The compilation of notes will be more extensive.

We solicit suggestions from interested users.

C. Jaschek and G. A. Wilkins (eds.), Compilation, Critical Evaluation, and Distribution of Stellar Data. 151.

PART IV

THE DISTRIBUTION OF DATA

DATA DISTRIBUTION

B. Hauck

Institut d'Astronomie de l'Université de Lausanne
et
Observatoire de Genève, Switzerland

I. Introduction

Before answering, or trying to answer, the question "how to distribute astronomical and astrophysical data", it is necessary to reply briefly to the question "why distribute these data?" This question is not so trivial as it first appears and it even hides the basic problem.

Data acquisition in itself is a single or manifold technique, often very delicate, constituting in itself a whole field of research. The interpretation of data follows it and will not necessarily be made by the same people. Very often only one possible aspect will be examined. It is consequently absolutely necessary that the data should be disseminated as widely as possible in order that they are used to the utmost. If it is necessary to assure the dissemination of data, it is no less necessary to ascertain their conservation. I shall give here only one example: the observations which have permitted to obtain the light curve of the quasar 3C273.

The necessity to distribute and to preserve the data being established, we can now examine the problems related to their distribution.

If a session of this colloquium is devoted to data distribution, it is because the traditional method of printing a catalogue or a list, and forwarding it to other observatories by some way or another is certainly no longer well suited to the present situation. On the one hand because the number of new data to be published increases at an extraordinarily high rate and, on the other

C. Jaschek and G. A. Wilkins (eds.), Compilation, Critical Evaluation, and Distribution of Stellar Data. 155-159.

hand, because there exist today other means.   It is also necessary to take into account the fact that it is often important to
have new data available very quickly and that some data have a
short lifetime.

II.   Increase of Data

C. Jaschek (1968, 1973) has already demonstrated the very
large increase, or inflation, of data in various fields of
stellar astronomy and it is not necessary for me to give once
more this information.   However, I shall give you briefly some
statistics concerning the photometric data.   These are given in
table 1.   The numbers given in this table appear to me to be a
good demonstration of the necessity to consider new methods for
data distribution.

III.   Methods of Distribution

In this section I shall examine essentially the question of
supports for the information.   It is clear that the classical
way - that of printing catalogues and lists of data - remains,
and will remain, a very useful and valid method.   Some journals
play an important part in this direction, in particular the
Astrophysical Journal and Astronomy and Astrophysics, in editing
Supplement Series intended for the dissemination of the data.
The increase of the latter brings about an increase in the number
of published pages.   The Supplement Series of Astronomy and Astrophysics contained 1075 pages in 1970 and 2979 in 1974!   This
leads also in many cases to an increased delay of publication.
It is still necessary to add some catalogues edited directly by
the observatories.

We may now ask ourselves whether this support is the only
one to maintain.   My answer is no.   It is indeed very useful to
have available on a shelf in one's own office catalogues which are
easily consulted and this fact must be taken into consideration.
However it is also necessary to bear in mind with regard to this
kind of distribution the fact that an up-to-date version is very
difficult to obtain and that the problem of the inflation of printed pages is more acute.   Furthermore, it is not possible to use
directly the data in this form with a computer.   This is certainly the greatest weakness of data distribution only in printed form.

Today it seems to me that it is essential that the basic support of catalogues should be magnetic tapes.   It will then be
possible to distribute them to people interested in using them in
some computations.   It is certainly easy to object that not everyone can work with magnetic tapes, but I wish to insist on the fact
that if I consider the magnetic tape as the best suitable basic

support for the information, I think nevertheless that it is not
the only means of distribution.   If you have a catalogue on magnetic tape, then you have at your disposal many various ways of
distribution, such as

(a)   a copy of the master tape
(b)   a listing (but this process is expensive)
(c)   offset copies
(d)   microfiches

   In the latter case, you certainly know of the existence of
machines which produce microfiches directly from a tape.

   Thus, a catalogue on magnetic tape cannot only be used with
a computer, but can also be distributed in numerous other ways.
To the use of the various supports mentioned, I wish to add the
possibility to use a catalogue on magnetic tape in the framework
of a system of remote computing, the user being linked to a central computer by only a console or through its own computer.
Such linkage of communications can exist either inside an observatory or on a national (or international) scale.   The example of
the INAG Computer network in France (Jung 1974) is a very good
one.

IV.   A Possible Solution for Distribution

   Let us now examine a possible solution for distribution which
will take simultaneously into account the wishes of some people
(quick access, if possible catalogue on library shelves) and of
others (possibility to treat the information directly on a computer).   It is also necessary to take into consideration the
problems of the editors and to avoid the congestion of journals!

   A catalogue must firstly be put by the authors on magnetic
tape, then it will be sent to the editor (with copies of listings)
of a journal publishing data.   It is important also that a catalogue be submitted to a referee.   In some cases, the catalogue
will be reproduced by offset, but in most cases a microfiche would
be made and attached to the issue containing a short description
of the catalogue.   Thus, the existence of the catalogue would be
known by all interested people.   Moreover, it would be possible
to give it an accurate reference in subsequent papers.   This is
a very important point for the authors.   The catalogue would also
have to go through the filter of a referee system, which is important if we wish to avoid a proliferation of bad catalogues.   As to
the magnetic tape, it would be sent after the microfiche process by
the editor to a data centre.   The centre is responsible for the
distribution of copies and the preservation of the master copy.
It is important that the astronomical community is guaranteed the

distribution and storage of the tapes, and only the collaboration
between a journal and a data centre can assure their distribution
and storage.

At the present time, an intermediate solution is proposed by
the Supplement Series of Astronomy and Astrophysics.   An author
can submit a short description of his catalogue.   Only this description and one page of the listing are printed; but the description and catalogue (in the form of a complete listing) are
examined by a referee.   The tape is deposited at the Stellar
Data Centre at Strasbourg.   This centre will store the master
copy and distribute copies, listings or microfiches.

V.   Data Centre and Data Distribution

During this paper we have seen that an important part is
devoted to data centres.   We can now discuss the links between
the various centres.   It seems to me out of the question to en-
visage only one centre involved with all astronomical and astrophysical data.   Such a centre would be rather overloaded and I
doubt its efficiency.   Now a data centre is useful only if it is
efficient, in particular by distributing data very rapidly.   I
return to a solution which I proposed some time ago (Hauck 1974).
A centre should be specialized in a very well-defined field
(Stellar Data Centre, for example).   It would have the responsibility of the keeping up-to-date, the distribution and the storage
of data connected with its activity.   In order to improve the
diffusion in defined geographic areas, some sub-centres would be
in charge of distribution in their own area.   They should also be
remotely accessible.   Their files would be regularly brought up-
to-date by the main centre responsible for the field concerned.
These sub-centres could act simultaneously as sub-centres for many
fields.

VI.   Conclusion

The present possibilities of diffusion and treatment of data
certainly give us a valuable advantage compared with our predeces-
sors.   It is possible to obtain up-to-date data regularly which
are easily accessible, on supports allowing various uses.   But it
is nevertheless necessary to have the willingness to use what is
at our disposal and to make sure that a coherent policy exists or
is applied between the various data centres (and sub-centres) together with close collaboration with specialized journals.

Table 1

I. *UBV* SYSTEM

Blanco, V.M. *et al.* (1970)	24582 entries
Mermilliod, J.-C. and Nicolet, B. (1976)	72943 entries

II. *uvbyβ* SYSTEM

Strömgren, B. and Perry, C.L. (1965)	1217 stars
Lindemann, E. and Hauck, B. (1973)	7603 stars
Hauck, B. and Mermilliod, M. (1975)	9407 stars
Hauck, B. and Mermilliod, M. (1976)	14589 stars

III. GENEVA SYSTEM

Rufener, F. *et al.*	(1964)	342 stars
Rufener, F. *et al.*	(1966)	686 stars
Rufener, F.	(1971)	1406 stars
Rufener, F.	(1976)	4670 stars

References

Blanco, V.M., Demers, S., Douglass, G.G. and FitzGerald, M.P.:
    1970, *Publ. U.S. Naval Obs.*, 2nd series, 21.
Hauck, B.: 1974, *Bulletin CDS*, Strasbourg, no. 6, 1.
Hauck, B. and Mermilliod, M.: 1975, *Astron. Astrophys. Suppl.*
    22, 235.
Jaschek, C.: 1968, *Publ. Astron. Soc. Pacific* 80, 654.
Jaschek, C.: 1973, in Ch. Fehrenbach and B.E. Westerlund (eds.),
    IAU Symp. 50, p. 275.
Jung, J.: 1974, *Bulletin CDS*, Strasbourg, no. 7, 2.
Lindemann, E. and Hauck, B.: 1973, *Astron. Astrophys. Suppl.*
    11, 119.
Mermilliod, J.-C. and Nicolet, B.: 1976, in preparation.
Rufener, F., Hauck, B., Goy, G., Peytremann, E. and Golay, M.:
    1964, *Publ. Obs. Genève*, série A, no. 66.
Rufener, F., Hauck, B., Goy, G., Peytremann, E. and Maeder, A.:
    1966, *J. Obs.* 49, 417.
Rufener, F.: 1971, *Astron. Astrophys.* 3, 181.
Rufener, F.: 1976, *Astron. Astrophys.* (in press).
Strömgren, B. and Perry, C.L.: 1965, Institute for Advanced
    Study, Princeton, no. J (second version, unpublished).

RETRIEVAL TECHNIQUES AND GRAPHICS DISPLAYS USING A COMPUTERIZED
STELLAR DATA BASE

Jaylee Mead
NASA/Goddard Space Flight Center

Theresa A. Nagy
Computer Sciences Corporation

ABSTRACT

A computerized astronomical data retrieval system, based on
the Goddard Cross Index of star catalogs and operable from a
remote terminal, has been developed. It permits retrieval of
stellar data as a function of the object's identification numbers,
descriptive parameters (magnitude and/or spectral type), or
position in the sky. In addition, software has been developed to
retrieve the full data entry from any of the eleven catalogs
currently included in the Goddard Cross Index, such as the Yale
Bright Star Catalog (YBS), the Boss General Catalog (GC), and
others---all in one computer run.

Four catalogs (Smithsonian Astrophysical Observatory Star
Catalog (SAO), The Revised New General Catalogue of Non-stellar
Astronomical Objects (RNGC), Reference Catalogue of Bright
Galaxies, and Two-Micron Sky Survey) have been sorted by Palomar
Sky Survey plate area and precessed to the epoch of the specific
plate. For any set of coordinates covered by the Palomar Survey
and the Whiteoak Extension, the computer can provide all the plate
numbers on which the position can be found. These plate areas
can be immediately accessed by computer; listings or plots to any
desired scale of any or all of the objects from the four catalogs
can be provided.

C. Jaschek and G. A. Wilkins (eds.), Compilation, Critical Evaluation, and Distribution of Stellar Data. 161-166.
Copyright © 1977 by D. Reidel Publishing Company, Dordrecht-Holland. All Rights Reserved.

I.   INTRODUCTION

During the past several years a large number of important stellar catalogs have been put into machine-readable form by various groups. A user can thus search by computer any of these star catalog tapes to obtain the data required for his observing program or theoretical study. To do this, he must know the catalog identification (ID) number such as Henry Draper (HD) or Durchmusterung (DM) number, or position in order to select objects from each catalog. He must know how the catalog is sequenced, for what epoch the positions are given, and the format of each record. He must be sure the tape is adapted to his computer and that it is blocked in an efficient manner for rapid processing of the data.

Becoming familiar with each star catalog tape can be a time-consuming and frustrating task. It would be much more efficient to give the computer a list of ID's in the system of any machine-readable catalog, or a position, or a range in spectral type and/ or magnitude, and obtain the available data from any or all of the catalogs. The development of such retrieval techniques and graphics displays using a computerized stellar data base at the NASA/Goddard Space Flight Center will be described in this paper.

II.   CAPABILITIES OF THE GODDARD DATA RETRIEVAL SYSTEM

The current Goddard data base consists of 28 machine-readable astronomical catalogs. Eleven of these catalogs have been combined into the Goddard Cross Index (GCI), which serves as the computer entry point to these catalogs. A more complete description of the GCI is given by Underhill, Mead and Nagy in this volume.

Below are some examples of how this data base and retrieval system are being used:

A.   Star catalog lookup or retrieval: A random set of HD numbers can be entered into the computer; they are first sorted by ascending HD. These HD's are then matched against HD's in the GCI, from which the YBS and GC numbers are pulled, where available. These YBS and GC ID's are then sorted in asending order and matched against each catalog (YBS and GC) to provide the data listing from the respective catalog for the original list of HD stars---all in one computer run.

B.   Preparation of candidates for observation by searching the data base for stars with given characteristics, such as location in a certain part of the sky, or having a brightness greater than a given magnitude, or falling within a given spectral type range. It is also possible to select stars on the basis of two or more characteristics, such as unreddened B stars, or high velocity binaries.

C. Identification of potential guide stars within a given field-
of-view, including specification of angular separation from
the target star plus magnitude and spectral class limitations.
This is especially useful for helping to locate and track
faint stars. It can also be used to anticipate bright stars
in or near the camera or telescope field-of-view, which may
cause problems due to scattered light (especially for spacecraft observations). We have found the Strasbourg <u>Catalogue
of Stellar Identifications</u> (CSI) to be especially useful in
this application since it contains over 400,000 stars.

D. Generation of plots of all catalog stars in or near the
telescope's field-of-view to scale of Palomar, other atlases,
or to the telescope itself for use as observing charts or to
aid in identifying unknown sources, such as x-ray sources.
Four catalogs (<u>Smithsonian Astrophysical Observatory Catalog</u>,
<u>The Revised New General Catalogue of Non-Stellar Astronomical
Objects</u>, <u>Reference Catalogue of Bright Galaxies</u>, and <u>Two-
Micron Sky Survey</u>) have been sorted by Palomar Sky Survey
plate area (Lund, J. and Dixon, R., 1973) and precessed to the
epoch of the specific plate, with x- and y- coordinates in mm
computed for each object. For any set of coordinates covered
by the Palomar Survey and the Whiteoak Extension, the computer
can provide all the plate numbers on which the position can
be found. These plate areas can be immediately accessed by
computer; listings or plots to any desired scale of any or
all of the objects from the four catalogs can be provided.
Figure 1 is a Calcomp plot of 3 types of objects appearing
on Palomar plate centered at $-18°$, 18^h06^m. Plotted are 360 SAO
stars (squares), 42 two-micron sky survey objects (diamonds)
and 15 RNGC objects (circles). Target circle in lower right
is 15' of arc in radius.

III. COOPERATION WITH NASA's NATIONAL SPACE SCIENCE DATA CENTER
(NSSDC)

The NSSDC, located at the Goddard Space Flight Center, is well-
equipped for retrieval and distribution of data to users. The
Laboratory for Optical Astronomy (LOA) group has worked closely with
the NSSDC in developing our existing data base. After checking
out a given machine-readable catalog, we deposit it with the NSSDC,
which then handles the distribution of that catalog to other users.
As soon as the first version of our computerized astronomical data
retrieval system is fully operational we plan to deposit this with
the NSSDC for processing of requests. As the Strasbourg Stellar
Data Center services the French astronomers through a computer
network centered at Meudon, is is expected the NSSDC will eventually
service North American astronomers via a telephone-computer hookup
at NASA/Goddard.

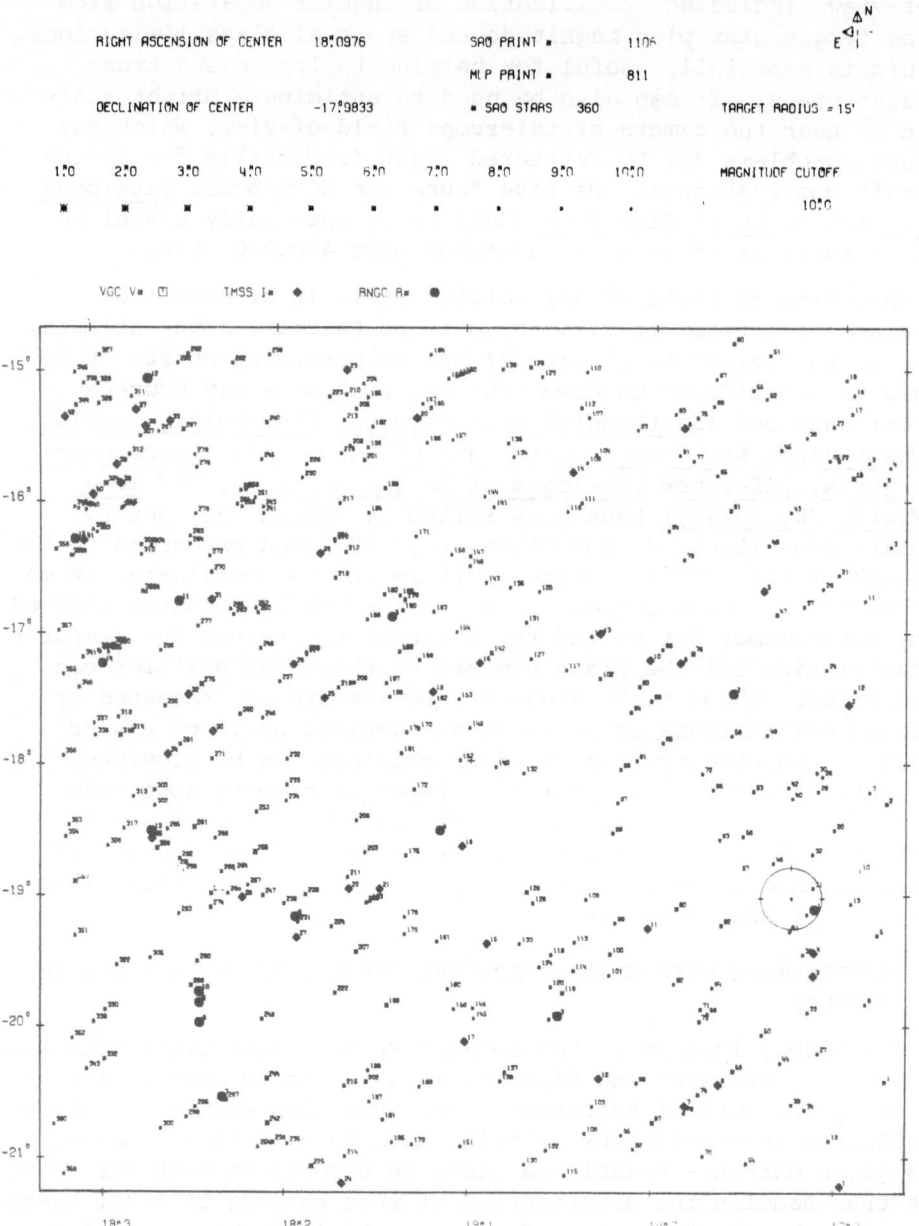

Figure 1 - Calcomp plot of 3 types of objects appearing
            on Palomar plate centered at -18°, 18ʰ06ᵐ.
            Plotted are 360 SAO stars (squares), 42 two-
            micron sky survey objects (diamonds) and 15
            RNGC objects (circles). Target circle in
            lower right is 15' of arc in radius.

IV. INTERAGENCY COORDINATING COMMITTEE FOR ASTRONOMY - DATA
     TASK FORCE

The Interagency Coordinating Committee for Astronomy, which
funds most of the astronomical research in the U.S., has requested,
through its Data Task Force, that the NSSDC survey the American
Astronomical Society membership to obtain information about
available compilations of astronomical data. The first survey
will be limited primarily to observational stellar and galactic
data. Solar system data, laboratory measurements, and model
results may be covered in later directories, if interest warrants
this. A computer-oriented questionnaire has been prepared for
mailing to anyone who wishes to describe his data files for
inclusion in the directory. Respondents are encouraged to submit
descriptions of both machine-readable and non-machine-readable
data in any stage of completion. The information obtained will be
keypunched for compilation into a fully indexed, loose-leaf
Directory of Astronomical Data Files. Supplements will be issued
to keep it current. The characteristics, or keywords, of each file
will be machine-tabulated to permit searches by computer. The
first edition will concentrate on North American data compilations.
The goal is to help a user locate data sources keyed to his
interests, along with enough descriptive information to permit him
to assess the value of the files for his use as well as the status
and availability of the compilations.

V. EXTENSION OF THE COMPUTERIZED DATA BASE

Appendix 1, Table of Nomenclature, from Kukarkin's Variable
Star Catalogue has been keypunched at Wellesley Observatory and
put on magnetic tape at Goddard. Since the Variable Star Catalog
does not include HD or DM numbers, this cross index by variable
star name and other ID's will be useful in providing definite
identifications for these stars and will greatly facilitate their
incorporation into our astronomical data base.

A magnetic tape version of the Cordoba Durchmusterung Catalog
is being prepared at the NSSDC. This should prove useful in
extending the completeness of the DM identifications in the
southern hemisphere.

VI. SUMMARY

Astronomers using observational data can realize extensive
benefits from the creation of a computerized astronomical data
retrieval system through the sharing of a data base with its
inherent expanded capabilities. Such an effort is being carried
out at the NASA/Goddard Space Flight Center and the NSSDC through
acquisition of star catalogs, development of retrieval techniques,
and creation of a Directory of Astronomical Data Files. A program
has been developed to retrieve with a single computer run the
full data entry from any of eleven catalogs currently included in

the Goddard Cross Index of star catalogs. Graphics displays, such as Calcomp overlay plots, are available for four catalogs, sorted by Palomar plate area.

The Strasbourg Stellar Data Center is making an invaluable contribution in the field of stellar studies by collecting, combining and distributing computerized stellar data. Their services have been extended not only to French and other European astronomers, but on an international level as well. It is our hope that the National Space Data Center at the Goddard Space Flight Center will soon provide similar services in the retrieval of stellar data for North American astronomers. Our further goal is that the Strasbourg and Goddard groups may work together even more closely in the future as we both serve the needs of international astronomy.

REFERENCE

Lund, J. and Dixon, R. 1973, Publ. Astr. Soc. Pac. 85, 230.

A MASTER LIST OF NON-STELLAR OBJECTS

Robert S. Dixon

Ohio State University Radio Observatory
and
Ohio State University Computer Center

It has been nearly 70 years since Dreyer completed his famous NGC and IC compendia of all the 13,000 non-stellar objects known in his time. Since then, the number of these objects known has increased by an order of magnitude, but this knowledge is scattered throughout the astronomical literature in such a way that it is nearly inaccessible from a practical standpoint to any individual.

The present work is an amalgamation of all known catalogs of non-stellar objects, in a uniform, easily readable form. It is intended not only for astronomers, but also for scientists and engineers in other fields who have need for rapid access to this basic reference data. The types of objects included are galaxies (including both normal and all specialized types such as interacting, peculiar, dwarf, Seyfert, etc.), clusters of galaxies, nebulae of all kinds (planetary, reflection, emission, absorption, etc.), blue objects, groups of stars (open and globular clusters, associations, rings, chains, etc.), quasi-stellar objects, supernovae and others. The information given for each object includes name, 1950.0 position, angular diameter in arcseconds, magnitude, description, and original reference. A portion of the work is shown in Figure 1. Approximately 185,000 listings appear in the full master list.

The specific purposes the work is intended to fulfill are:
1. To provide a moderate amount of information about every known non-stellar object, recognizing that it is not possible in a single work to include all known information about all known objects.
2. To serve as a pointer to the astronomical literature, for deeper study. Complete references are provided for this purpose.
3. To enable cross-comparisons among different original catalogs to be done, both among objects of the same type and among objects of different types.

C. Jaschek and G. A. Wilkins (eds.), Compilation, Critical Evaluation, and Distribution of Stellar Data. 167-169.

OBJECT NAME	RIGHT ASCN.	DECLINATION	DIAM.	MAGN.	TYPE OF OBJECT
PHL 2639	00 03 48.	- 18 48		18.4	BLUE STELLAR OBJECT
PHL 6277	00 03 48.	- 22 08		17.1	BLUE STELLAR OBJECT
TON-S 0139	00 03 48.	- 22 09		15.1	BLUE STAR
LB 04974	00 03 48.	- 29 35		17.4	FAINT BLUE STAR
LB 04975	00 03 48.	- 29 43		18.9	FAINT BLUE STAR
LB 04976	00 03 48.	- 30 35		17.4	FAINT BLUE STAR
AGU 01	00 03 48.	- 41 45 00.		12.5	2 INTERACTING GALAXIES
LB 01542	00 03 48.	- 48 03		14.3	FAINT BLUE STAR
BC 3CR2	00 03 48.70	- 00 21 06.6		19.35	QUASI-STELLAR OBJECT
RNGC 7829	00 03 49.	- 13 41		14.0	GALAXY
RNGC 7828	00 03 49.	- 13 41		14.0	GALAXY
PHL 6279	00 03 54.	- 06 14		18.1	BLUE STELLAR OBJECT
LB 04977	00 03 54.	- 30 23		19.2	FAINT BLUE STAR
LB 01543	00 03 54.	- 46 31		15.1	FAINT BLUE STAR
RNGC 7833	00 03 55.	+ 27 22			NON-EXISTENT OBJECT
HOLM 001B	00 03 58.	+ 04 51	12	14.8	PART OF MULTIPLE GALAXY
LB 04002	00 03 58.	+ 29 56 06.		17.7	FAINT BLUE STAR
ARC 2708	00 03 58.	- 17 12		17.4	RICH CLUSTER OF GALAXIES
LBN 0585	00 04 00.	+ 65 21	60		BRIGHT NEBULA
PEL 0664	00 04 00.	+ 07 56		17.1	BLUE STELLAR OBJECT
UGC 00047	00 04 00.	+ 17 00	84	17.	GALAXY
ZC 0004-0+2446	00 04 00.	+ 24 46	1810		CLUSTER OF GALAXIES
UGC 00048	00 04 00.	+ 47 36	126	16.0	GALAXY SB C
ISS 0001	00 04 00.	+ 63 31	136		STELLAR RING
LDN 1275	00 04 00.	+ 67 10	540		DARK NEBULA

Figure 1 - A sample page from the master list.

4. To make all the original catalogs available in computer-readable form. This form generally includes not only the information included in this work, but often whatever other information the original author provided.

A special attempt has been made to make this work as self-explanatory and uncluttered as possible. Abbreviations and special symbols have been avoided. Consistent and simple units of measure have been used, and no subsidiary tables or detailed explanation are required to understand and utilize all of the basic data presented. No new nomenclature has been introduced; that assigned by the original author has been retained.

This master list will be made available in both book and magnetic tape form. The magnetic tape form will be updated indefinitely to include all new catalogs that are published, and to incorporate errata.

PROPOSAL FOR A DATA CENTRE ON GALACTIC NON-STELLAR OBJECTS

M. C. Lortet

Observatoire de Meudon, France.

Bibliographic work was begun in 1969 on Interstellar Matter and Galactic Non-Stellar Objects, mainly those found in regions suspected to be favourable to star formation.

We are contemplating the systematization and automatization of this work, taking advantage of the tools now worked out for stellar data at the Centre de Données Stellaires de Strasbourg.

I. PROPOSALS

1. Centralization of the bibliography

The bibliography might be taken primarily from our file on Interstellar Matter, and might also be collected by asking the authors of extensive studies (Catalogues, Surveys) to send their work directly to the Centre, as soon as it is ready to be submitted for publication.

2. Analysis of data

Each paper, in addition to the objects there described, should be codified with appropriate key-words. The key-words would be chosen by the authors and/or by the analytical working group, in view of subsequent work.

Such key-words might be:
. theoretical model, describing a specified individual object.

C. Jaschek and G. A. Wilkins (eds.), Compilation, Critical Evaluation, and Distribution of Stellar Data. 171-173.

. maps or photographs.
. optical line-intensity ratios.
. kinematic studies (velocity field, lines profiles, kinematic distance).
. continuum radio studies (for instance a new field will be
  millimeter continuum radio studies).

Subsequent work might be done by separate working groups,
from the documentation sorted by the computer, for instance:
. bibliography on a specified object or sample of objects
  (as is currently done for stars by the Centre de Données
  Stellaires de Strasbourg).
. study of a specified parameter (or class of parameters)
  for a sample of objects (for instance, bibliography on
  infrared emission of H II regions, detection of molecular
  lines emission in different kinds of galactic non-stellar
  objects).
. preparation of catalogues with well arranged bibliography
  of a sample of objects (for instance a catalogue, a study,
  and a bibliography of a complex H II, H I, molecular cloud
  region where star formation is going on).

II. PROBLEMS

1. Choice of key-words

We cannot attempt to make an exhaustive codification: it
would be too time-consuming at the input and our experience is
that it would be totally inefficient at the output.

Instead we devised a limited number of key-words, not too
restrictive in their meaning, and most of them connected with
current research in France, falling into the following cate-
gories:
. kind of information found in the paper (new observation,
  theoretical model, laboratory physics, review paper).
. kind of instrumentation and wavelength.
. type of objects studied (H II region, dark cloud).
. physical parameters studied (line-intensity ratios, line
  profile, dust content).

We propose to add a set of more specific key-words (maybe
subcategories of the preceding ones), devised in view of well
defined subsequent work (paragraph I, 2).

2. Choice of the papers to be collected

The introduction of key-words may allow inclusion of theoretical papers, and not only those devoted to one or several real
objects.

3. Cross-identification catalogue for non-stellar objects

A compilation of some synonymies or inclusions may be
quickly carried out from existing Catalogues. However, we have
to build a more systematic and updated Catalogue, a problem far
more complex than for stellar objects or planetary nebulae. As a
guiding principle, we possibly should give the priority to two
extreme kinds of designations, one very precise for location (to
within a few arc seconds), the other indicating into which larger
entity (cloud, large H II region) the object can be inserted in
its projection on the sky.

MAIN FEATURES OF THE STELLAR BIBLIOGRAPHIC FILE

F. Ochsenbein, F. Spite
Observatoire de Strasbourg, Observatoire de Paris-Meudon

Summary : Due to the increasing amount of the literature about
stellar studies, the gathering of the published information
about a given star becomes a tedious and time consuming task.
For enabling an automatic retrieval of information, a bibliographic file has been built in machine readable form through
collaboration between Paris and Strasbourg Observatories. The
file begins with the year 1950, and is kept up to date. Knowing
the name of a star, a program prints the titles of the papers
quoting this star. However, only the main journals were covered
at first, and some other limitations have been imposed. The first
part of this file will be very soon published on microfiches.
Meanwhile the information can be obtained from the Strasbourg
Stellar Data Center, (Centre de Données Stellaires).

It is a tedious and time-consuming task to gather in the
literature the information about a star or stellar object.
Prof. R. Cayrel initiated in Paris-Meudon Observatory a bibliographic file, organised in a way suitable for a further automatic
retrieval of information. The astronomical and astrophysical
periodicals are read (not all of them, we will discuss this point
further) and the papers quoting names of individual stars are
looked for. For such a paper, the title and bibliographic reference is punched,on one hand, the list of quoted stars on the
other hand. At a later stage, planetary nebulae have been added
to the file.

Through collaboration with the Stellar Data Center in
Strasbourg, these informations are put in the memory of a computer in the proper way, and a set of programmes has been built

to process these informations. Especially it is possible, knowing
the name of a star, or stellar object,to get the titles of the
papers quoting this star or object. The situation is not however
as satisfactory as it may seem, because several limitations
occur:

1) Up to now, most of the cross-references have been built
for the names of the stars, and it may be that the complete
solution of this problem will never be reached for faint stars
(m_v >9) due to the continuous appearance of new designations of
stars and stellar objects. So that, for these fainter stars,you
have to enumerate all the names of your star that you can think of.

2) The bibliographic file begins for the year 1950 and is kept
up to date with about one year time lag.

3) The lists including more than 1000 objects are excluded from
the file. They are considered as small catalogues by themselves,
and can be easily searched. These small catalogues are listed and
are not very numerous (about 30 for the years 1950 to 1975).

4) Of course not all astronomical periodicals are covered. For
the years 1950-1972, only 12 periodicals were included. Starting
with the literature of 1973, more than 30 periodicals are covered.

5) Starting with 1973, discrete X-ray sources and infra-red
sources are included.

6) Stars numbered in clusters or special fields, without names
in classical catalogues are not included. Faint stars without coordinates are not included.

7) The bibliographic file is not error-free. Any user should
be ready to cope with errors and omissions. We ask to any user
finding errors or omissions to mention them to us.

8) Finally, the use of the bibliographic file is not instantaneous. Taking into account free days and vacations, and mail
delay, the process is a rather slow one for people who are not in
France.

As a solution to this delay, Prof. Jaschek had the idea of
publishing this file as a "Bibliographical Star Index". The part
covering the years 1950 to 1972 will be published on microfiches
next week. This Index is arranged by star names, and a preface
describing the main features of the Index is joined to the microfiche edition (Cayrel, 1976). Let us recall here that in the
bibliographical Star Index the code of the references (including
the year) are listed in front of each star name ; the references
are listed separately at the end, as it is done in Jaschek's well

known catalogue of stellar spectra (1976). This would enable to
get at once the references of the papers quoting a star between
1950 and 1972 in the 12 periodicals included at this time.

The next edition, dealing with the years 1973 and 1974 will
be available at the end of 1976.

It is planned to issue annually a microfiche edition of the
stellar bibliography.

Finally, a few data about the bibliographic file may help to
get an idea of the amount of data gathered :

Years 1950-1972 : 6 658 references and 56 000 different
objects ; on the average, each star has 3.4 references.

Year 1973 : 648 references, 13 512 objects.
The list of planetary nebulae included about 1 200 objects
and 7 references per object.

Many people have worked on this rather tedious and painstaking task : Prof. C. Jaschek, F. Ochsenbein, Drs. J. Jung,
A. Valbousquet, Mrs. Bischoff and Mrs. Wagner who has done most of
the punching work in Strasbourg; Prof. R. Cayrel, Mrs. Kirchner
and Dr. F. Spite in Meudon. The work about planetary nebulae has
been done by Dr. A. Acker and Mr. Marcout in Strasbourg.

The bibliographic file is frequently used by French astronomers,
and it is hoped that this work will be useful to the astronomical
community.

BIBLIOGRAPHY

Cayrel, R., Spite, F., Jung, J., Kirchner, S., Ochsenbein, F.,
Valbousquet, A. 1976, Bibliographical Star Index, Microfiche
Edition, Stellar Data Center, Strasbourg.

Jaschek, C., Conde, F., de Sierra, A. 1964, Catalogue of stellar
Spectra Classified in the Morgan-Keenan System, La Plata.

THE VISUAL DOUBLE STAR CATALOGUES

Charles E. Worley

U.S. Naval Observatory

ABSTRACT

The historical development and current status of the two
visual double star catalogues maintained at the Naval Observatory are discussed.

The Naval Observatory maintains two visual double star
catalogues by international agreement. These catalogues are
among the oldest continuously-updated data files in astronomy,
having been established more than a century ago. The first of
these catalogues, called the Index Catalogue, lists all known
double and multiple stars; it currently contains 70,295 entries.
The second catalogue is called the Observation Catalogue; it
lists all observations of double stars published since 1927,
plus a considerable number of earlier measures: entries total
301,995 as of 1976.5. Throughout their history these catalogues have exercised immense influence on the development of
double star astronomy, while at the same time proving to be of
great value to the general astronomical community. Their value
has been enhanced because they always have been maintained by
experts in the field, and because there has been periodic publication of the accumulated data.

S. W. Burnham began to collect double star data about 1870,
and in 1906 he published A General Catalogue of Double Stars
within 121° of the North Pole. This catalogue, usually abbre-
viated BDS, consists of two sections. The first is an index
catalogue, which lists positions and other pertinent identifica-

C. Jaschek and G. A. Wilkins (eds.), Compilation, Critical Evaluation, and Distribution of Stellar Data. 179-182.
Copyright © 1977 by D. Reidel Publishing Company, Dordrecht-Holland. All Rights Reserved.

tion information for all double stars known in 1903. Entries
total 13,665. The second, and much larger, portion of the catalogue gives a selected list of measures of each pair, plus
references to all omitted measures, and also includes extensive
notes on many stars.

The high rate of double star discovery and observation in
the first decades of this century soon made Burnham's catalogue
obsolete. Burnham himself continued the collection of new data
until 1912, when Eric Doolittle assumed this duty, which he
continued until his death in 1920. R. G. Aitken then began his
long association with the catalogues, which did not end until
after World War II. In 1932 Aitken published his New General
Catalogue of Double Stars within 120° of the North Pole (ADS).
This catalogue rejected about a third of the wide pairs listed
by Burnham, yet nevertheless contained 17,180 double stars.
Unlike the BDS, the ADS attempted to list all measures made of
each object between 1903 and 1927; however, space limitations
forced Aitken to combine individual measures into means in many
instances.

Southern double stars were cared for by R. T. A. Innes,
who produced a Reference Catalogue as early as 1899. His major
catalogue, however, was the Southern Double Star Catalogue
(SDS), published in loose-leaf form in 1926-27. This work also
omitted wide pairs, but still contained nearly 10,000 objects.

Following the retirements of Aitken and Innes, H. M.
Jeffers became responsible for the northern doubles, and W. H.
van den Bos for those south of -20°. About twenty-five years
ago the transfer of the northern data from hand-written cards
to punch cards was begun, and, following the later decision to
combine northern and southern catalogues, the southern material
was also punched. A comprehensive, but not complete, catalogue
of observations was thus formed for the first time, and from it
the Index Catalogue of Visual Double Stars, 1961.0 (IDS) was
constructed and published in 1963 by Jeffers, van den Bos, and
Greeby. This catalogue included all the pairs omitted by
Aitken and Innes, but gave no individual measures. Objects
listed totalled 64,237. A serious defect, in my opinion, was
the omission of a bibliography of the published measures, which
formed such a helpful feature of the BDS and the ADS.

In 1964 the Naval Observatory assumed responsibility for
the two catalogues. Since that time 6,048 double stars have
been added to the Index Catalogue, while the Observation Catalogue has grown by 93,123 cards. For some time now a project has
been underway to complete the Observation Catalogue by the
addition of all of the pre-1927 measures. Of the total of
301,995 cards now filed in this catalogue, 46,085 represent these

older observations. Another 30,000 cards are in preparation.
My best estimate is that the older material will not substantially exceed 150,000 cards, so that roughly half of this project is finished.

Responsibility for the catalogues also extends to error detection and correction. About 10,000 corrections have been incorporated in the last eleven years. Many of these corrections,
however, represent only minor alterations of format designed to
increase the homogeneity of the data. The Naval Observatory has
also agreed to supply three depositories with exact copies of
all additions and alterations to the catalogues. The three
depositories are currently at the Lick, Nice, and Royal
Greenwich Observatories, to which card shipments are made on
an annual basis.

Users of visual double star data fall rather naturally into two groups. The non-specialist usually is interested in
knowing if certain stars in an observation list or catalogue
are double; and, if so whether or not the pairs are physical.
Consequently, we often receive requests for data on large numbers of stars from such users. On the other hand, the specialist usually wants extensive data on relatively few objects,
and thus it is generally possible to serve his needs more quickly. To date, about 250 data requests have been filled, ex-
cluding telephone inquiries and our own extensive internal use
of this material. Requests have ranged from data on a single
star to more than 6,000 objects. The entire Index Catalogue has
been requested and supplied on tape perhaps a dozen times. Full
documentation is included with each tape, and with the listings
we provide a complete description of coded data plus an explicit
reference to the publication in which the data originally appeared. Small quantities of data are supplied free, but for
extensive requests we ask for compensation in the form of blank
cards or tapes.

The major problem we have experienced with users of the
double star data is that of the form of their request. Since
the catalogues are ordered by 1900 coordinates, we must have
this information in order to remove the desired data from the
catalogue. Because of staff limitations, we find it impossible
to accommodate users who send requests using epochs other than
1900, or other identifications, unless their request happens
to be for a dozen objects or less. In the case of tape users,
we are able to supply either 7 or 9-track, with a number of
choices of recording density.

In conclusion, experience makes it evident that availability of the double star catalogues in machine-readable form fills
a wide and growing need in the astronomical community. Comple-

tion of the data file through inclusion of the older material,
plus the continuing correction and homogenization process, give
these catalogues steadily increasing value. While the creation
and maintenance of catalogues is hard and unexciting work, the
history of our science amply demonstrates the permanent value
of such endeavors.

SYSTEMATIC DIFFERENCES IN TRIGONOMETRIC PARALLAXES FROM DIFFERENT OBSERVATORIES

W.F.van Altena, E.D.Hoffleit and H.A.Smith

Yale University Observatory

ABSTRACT

Systematic differences in trigonometric parallaxes between Allegheny Observatory and Yale Observatory, between Allegheny Observatory and McCormick Observatory and between the Cape Observatory and Yale Observatory have been investigated for stars common to each pair. The differences found correlate with right ascension, naturally suggesting some sort of annual influence. It is proposed that these differences are related to differences in the annual temperature cycle between observatories, possibly through the mechanism of temperature dependent decentering of the telescope objectives. A dependence upon spectral type was also discovered in the differences between the relative parallaxes from Allegheny and from Yale. Further work is needed to clarify the nature of these systematic effects and to insure that they do not significantly bias available trigonometric parallaxes.

It is proposed that a new parallax catalogue be constructed at Yale after a thorough statistical analysis of all available trigonometric parallaxes has been made. We solicit suggestions and recommendations from interested users.

I. INTRODUCTION

It has long been remarked that trigonometric parallaxes derived for stars common to the programs of several observatories show small systematic differences from one observatory to another. Particularly pronounced are the differences between parallaxes from northern hemisphere observatories and from those of the southern hemisphere, which amount to about $0\overset{\prime\prime}{.}005$ (Strand, 1971). Though

C. Jaschek and G. A. Wilkins (eds.), Compilation, Critical Evaluation, and Distribution of Stellar Data. 183-189.
Copyright © 1977 by D. Reidel Publishing Company, Dordrecht-Holland. All Rights Reserved.

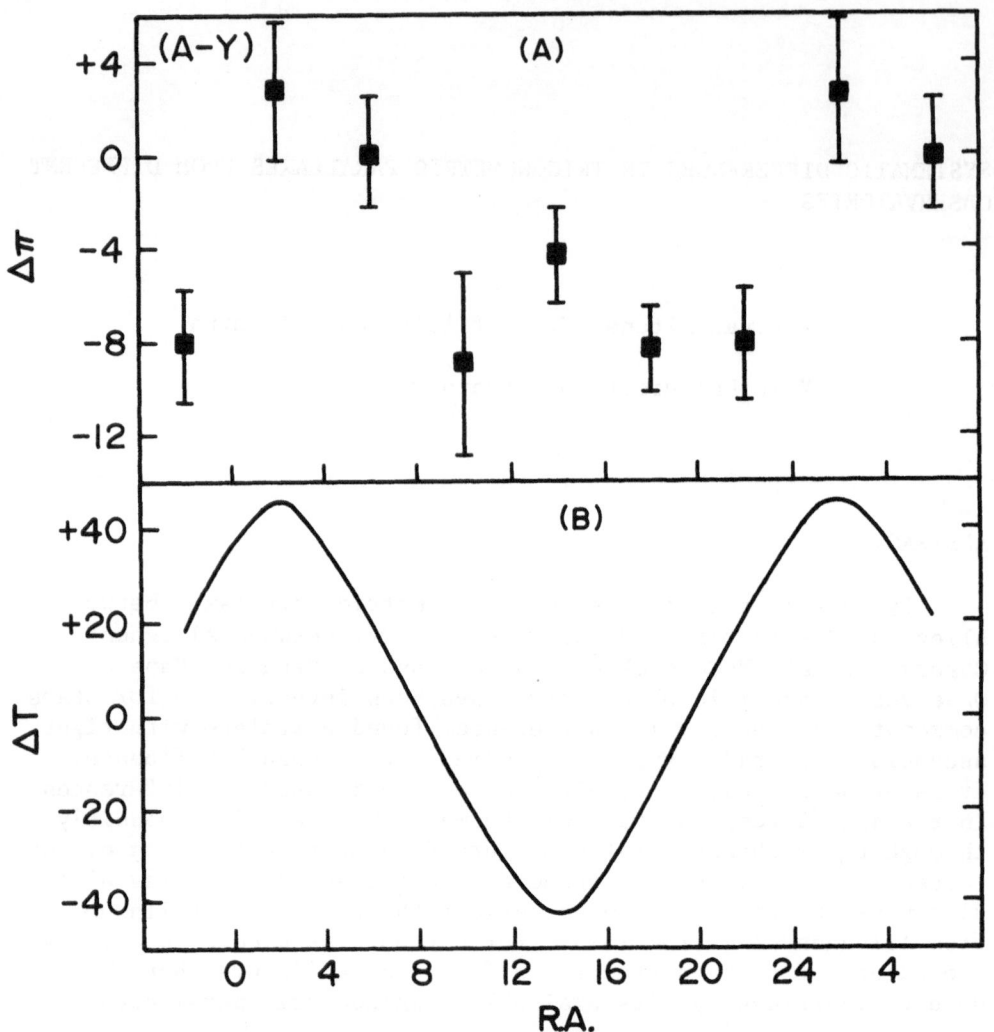

Figure 1.   (A) The (A-Y) parallax differences in units of 0.001
            arc-sec as a function of the right ascension. (B)
            The difference in temperature range, evening minus
            morning, between Pittsburg and Johannesburg as a
            function of the right ascension.

some proposals have been advanced (e.g. Atkinson, 1971) the cause
of these differences has yet to be established. This paper explores
in a preliminary manner some of the systematic effects between
parallaxes from two northern sites, Allegheny and McCormick and
from two southern sites, Yale and the Cape.

    We also propose to construct a new Parallax Catalogue after a
thorough statistical analysis of all available trigonometric
parallaxes. We solicit suggestions from interested users.

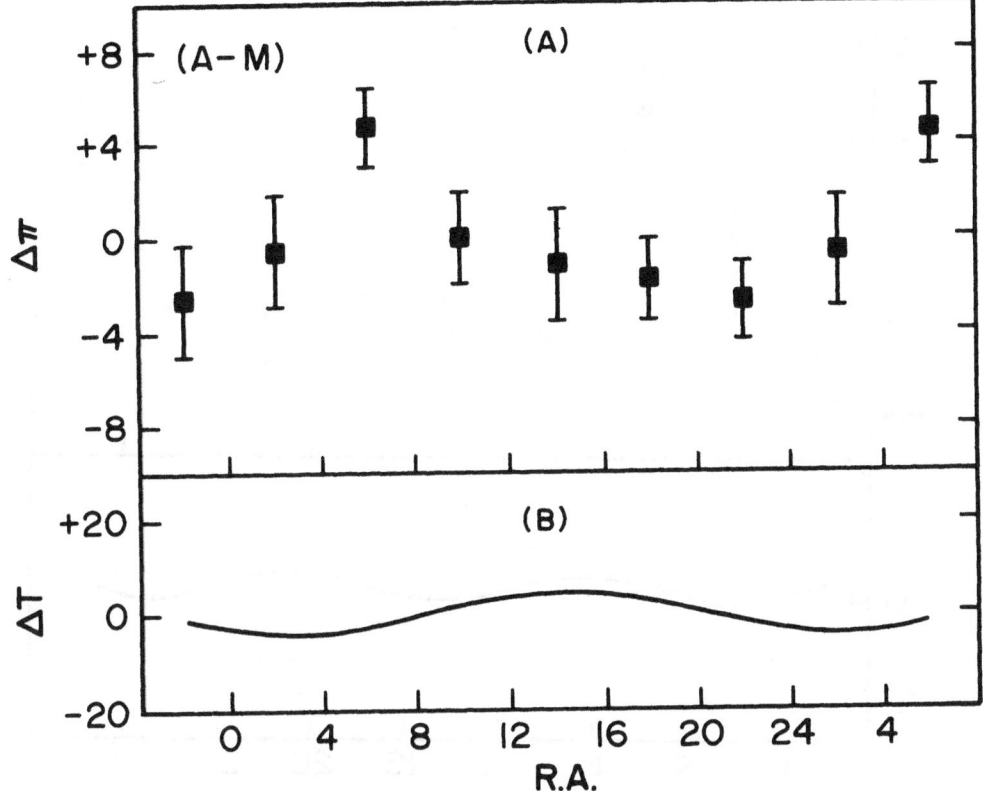

Figure 2.   (A) The (A-M) parallax differences in units of 0.001
arc-sec as a function of the right ascension. (B)
The difference in temperture range, evening minus
morning, between Pittsburg and Richmond as a function
of the right ascension.

II.   DATA

The data for these comparisons are the relative parallaxes for
295 stars common to the programs of Allegheny and Yale, 499 stars
common to the Cape and Yale programs, and for 680 stars common to the
Allegheny and McCormick progroms. These were obtained from the
General Catalogue of Trigonometric Stellar Parallaxes (Jenkins,1952).
Differences were computed in the sense Allegheny minus Yale, Cape
minus Yale and Allegheny minus McCormick from the relative parallaxes
of the common stars.

III.   RESULTS

The mean differences were, Allegheny minus Yale, -0.005, Cape
minus Yale, -0.002, and Allegheny minus McCormick, -0.001. An
interesting result is obtained, however, if instead of looking at

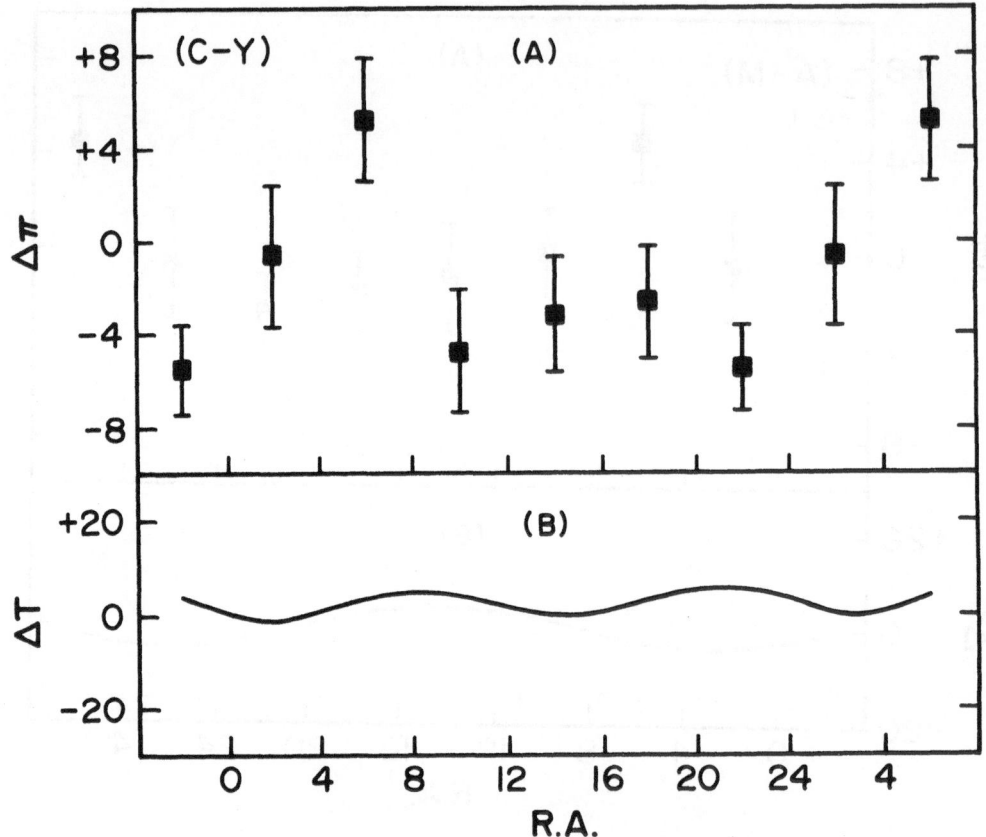

Figure 3.   (A) The (C-Y) parallax differences in units of 0.001
            arc-sec as a function of the right ascension. (B)
            The difference in temperature range, evening minus
            morning, between Capetown and Johannesburg as a
            function of the right ascension.

the mean difference of all of the parallaxes we divide the stars
into groups by right ascension. Figures 1a, 2a, and 3a plot the
differences (A-Y), (A-M) and (C-Y) for the stars in bins spanning
four hour zones of right ascension. The differences (A-Y), (C-Y)
and perhaps (A-M) seem to correlate with right ascension. Similar
systematic errors in right ascension have been pointed out by van
Maanen (1933), Mitchell (1934), Davidson (1934), Sterne (1935),
Dahlgren (1960), and Ljunggren and Oja (1965).

     Since the photographic plates from which these parallaxes were
determined were taken near the meridian the association of mean
difference with right ascension suggests an annual effect. Some
sort of optical effect must be present, more likely in the telescope
than in the atmosphere since we are dealing with parallax differences,
and moreover, with x parallaxes which should not be especially

sensitive to seasonal changes of the air. A plausible seasonal
influence is the annual cycle of temperature variation (Davidson
1934). Evening and morning plates are taken at different times
of the year for a given star, and consequently at different
average temperatures and temperature gradients. It is conceivable
that this temperature variation could alter the optical properties
of the telescope (see for example, Ianna[1965] and Kamper [1971])
between the taking of morning and evening plates for each bin of
right ascension.

To test this hypothesis mean monthly temperatures for Pittsburg
(Allegheny), Richmond (McCormick), Johannesburg (Yale) and Capetown
(Cape) were obtained from Nelson (1968). Estimates were then made
of the range of temperature between the taking of the evening and
morning plates for each bin of right ascension and for each
observatory. Because we are interested in differential effects,
the differences between the ranges for Allegheny and Yale, Allegheny
and McCormick, and Cape and Yale were calculated. These are plotted
in Figures 1b, 2b, and 3b. Comparing Figures 1a and 1b we find
that they correlate suggestively.

Although the comparison of Figs. 1a and 1b is suggestive,
similar but small parallax differences in the (A-M) and (C-Y) data
exist even when the temperature range variations are in phase and
therefore nearly cancel (Figs 2b and 3b). This is perhaps not too
surprising since we are dealing with four different telescopes
whose objectives may react in different ways to thermal gradients
or to different temperatures.

IV. DISCUSSION

There is some evidence, then, for a systematic annual error in
trigonometric parallax determinations resulting from the annual
temperature cycle. It is more difficult to establish the mechanism
by which this error enters. Some nonlinear effect must be at fault,
else the error would have been removed during the linear depend-
encies reduction of the measurements. Atkinson (1971) has proposed
a possible culprit: temperature dependent decentering of the
components of the telescope objective. Conrady (1919) has studied
the aberrations expected from such decentering, finding that extra-
axial points shift with respect to field center in a nonlinear way.

While this mechanism is plausible, it cannot be established by
the present results. The actual detection of decentering aberrations
must await experimentation such as that advocated by Atkinson. However, other consequences of decentering can be looked for in the
parallaxes. Davidson (1934) has pointed out that inconstant decentering might introduce a systematic error depending upon color.
Figure 4 illustrates the differences in the relative parallaxes,

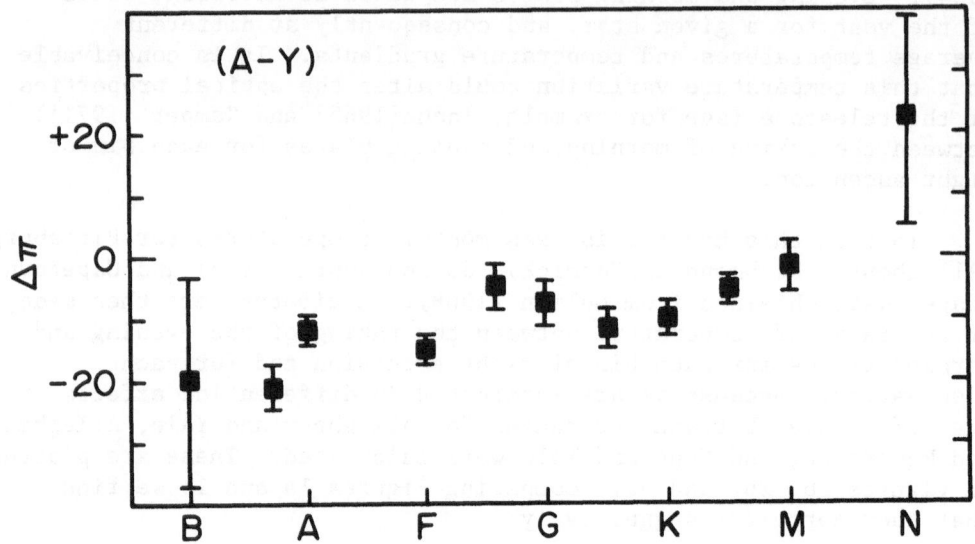

Figure 4.   The (A-Y) parallax differences in units of 0.001
            arc-sec as a function of the spectral type.

Allegheny minus Yale, as a function of the spectral type of the
parallax star. Despite considerable scatter a trend is evident,
the mean difference for M stars being about 0.''01 greater than for
AO stars. Again, it is possible, but by no means certain, that
this results from temperature dependent decentering.

     In a recent investigation, Kamper (in preparation) finds a
systematic difference in the astrometric data obtained with the
Yale refractor between its Johannesburg and Mt. Stromlo locations.
He notes that there is reason to believe that the refractor was
misaligned during part of the time while it was at Johannesburg
thereby producing color dependent shifts in stellar positions.
Turon LaCarrieu and Creze (1976) also comment that they feel that
the southern parallaxes are systematically in error.

     There is thus a case for re-examining the systematic differences
between parallaxes from different observatories in greater detail.
Though Gliese (1972) has suggested that the Allegheny system may
on the whole be free from substantial systematic errors, this
result needs confirmation. Since the errors may vary with right
ascension it is possible that certain parts of the sky may be
subject to systematic effects much larger than that of the entire
Allegheny system. There can be little complacency until it is

reliably established that these systematic errors do not translate into significant errors in the luminosities and masses of stars.

REFERENCES

Atkinson, R. d'A. (1971). Publ. Leander McCormick Obs. XVI,259.
Conrady, A.E. (1919). Mon. Not. Roy. Astr. Soc. 79, 384.
Dahlgren, T. (1960). Ark.Astr. 2,No.45.
Davidson, C.R. (1934). Obs. 57, 236.
Gliese, W. (1972). Quar.Journ.Roy.Astr. Soc. 13, 138.
Ianna, P.A.(1965). Vistas in Astron. 6, 94.
Jenkins, L.F. (1952). General Catalog of Trigonometric Stellar
    Parallaxes (Yale University).
Kamper, K.W. (1971). Publ. Leander McCormick Obs. XVI,285.
Ljunggren, B., and Oja,T. (1965). Ark. Astr. 3, No.35.
van Maanen, A. (1933). Astrophys.J. 78, 189.
Mitchell, S.A. (1934). Astrophys. J. 80, 200.
Nelson, H.L. (1968). Climatic Data for Representative Stations
    of the World, (Univ. of Nebraska Press).
Sterne, T.E. (1975). Astrophys.J. 81, 45.
Strand, K. Aa. (1971). Publ. Leander McCormick Obs.,XVI,15.
Turon LaCarrieu, C., and Creze, M. (1976). (preprint).

THE ASTRONOMICAL DATA SYSTEMS GROUP IN JAPAN

Yoichi Terashita

Computer Center, Kanazawa Institute of Technology
Kanazawa 921, Japan

1. INTRODUCTION

In spite of the relatively large size of the astronomical
community in Japan, there have been rather few efforts to
organize an astronomical data system(s) and to take advantage of
such systems. As a matter of fact, it has been realized that
astronomical data, as the volume grows, are in many cases easier
to handle in computer readable form than in printed form, and
there is a growing trend of recording observational data or large
tables on magnetic tapes and the likes, and of acquiring such data
from other investigators (notably from investigators in other
lands). However, it has not always been realized that machine-
readable data bring about various kinds of problems when they are
to be used by a number of astronomers whose fields of interest
are different from each other or who works at geometrically
scattered insitutions.

Concerned about such problems associated with the use of
astronomical data in computer-readable form, a small group,
including the author, held informal colloquia first in 1973 and
again in 1974. Through these colloquia it was found that not
small a number of astronomers are concerned about an organized
approach for the computer-readable astronomical data. The major
reasons for the difficulties arising from the use of such data on
a common basis were summerized as follows.

(a) Schemes for representing data on, say, magnetic tapes
are not as apparent as in the case of data in printed
form.

(b) In order to process the data, one must have an access
    to a computer system offering the capabilities of
    flexible tape/disk operations. However, such is not
    always the case at universities in Japan.
(c) The fact that computers of various kinds are used at
    different institutions causes a compatibility problem.
(d) There is no library service available for astronomical
    data tapes. An alternative would be for every
    investigator to own his copies covering a wide range of
    research fields, but it is of course not practical.

In 1975 the Ministry of Education, Science and Culture
announced the plan for sponsoring a three-year research project
on the organization of scientific information. Our proposal
concerning astronomical data systems was approved, and the whole
project started in the spring of this year. The project is one
of the so-called 'special research projects', in which the
emphasis is placed on coordinated research activities among their
member groups. This particular project is entitled 'Formation
Process of Information Systems and Organization of Scientific
Information', and consists of fifty seven research groups.
About two thirds of these groups are engaged in the developments
of data processing methodologies, such as input technologies,
pattern recognition, language analysis, and data base management.
The remaining groups are mostly engaged in the application of
such methodologies to specific fields of science, including,
beside astronomy, many branches of chemistry and related fields,
nuclear physics, mathematics, geology, etc.. A few group from
humanities are also participating. It should be noted that many
of the 'applications groups' are concerned with technical
documents. In the case of astronomy, however, we feel it is more
important and urgent to establish an information system for
observational data than do the same for astronomical documents,
at least for the time being.

2. GROUP ORGANIZATION

The astronomy group within the above mentioned project
consists of six colleagues. They are,

Dr. Kiichiro Furukawa and Dr. Shiro Nishimura
    of the Tokyo Astronomical Observatory,
Dr. Akira Uesugi
    of the Department of Astronomy, University of Kyoto,
Dr. Yoichi Terashita, Mr. Ichiro Fukuda and
Dr. Takashi Kusaka
    of the Computer Center, Kanazawa Institute of
    Technology.

The research fields of the members include position astronomy, stellar atmospheres, stellar evolution, and computer science. Participations of colleagues in other fields such as galaxies and radio astronomy are highly desired.

3. FACILITIES

Data processing facilities available to the members of the group are an IBM 370/158-II system (Kanazawa Institute of Technology), a FACOM 230/58 system (Tokyo Astronomical Observatory), and a FACOM M-190 system (University of Kyoto). Under present circumstances, the Kanazawa system is thought to be most fitted for computer-related operations involving extensive file processing. It is also equipped with an interactive terminal system and a key-punch service.

As for the data processing facilities at large, large-scale computer systems are available at major universities, and they can be used by scientists at other institutions at relatively low rates. Users can bring (or mail) their input decks or use time-sharing (or remote job entry) terminals where such things are available. Although these systems are large and fast, they are primarily designed to handle large volumes of numerical computation jobs, and present some inconveniences to jobs requiring extensive file operations. A long-term project of connecting the major systems is in progress. When the project is completed, the resulting computer network is expected to have an enormous impact on scientific information systems.

4. ACTIVITIES

After having several meetings within the group it has been agreed upon that a proto-type astronomical data system should be established and be evaluated within the group, allowing for an extended service for colleagues who might be interested in using such a system. By an astronomical data system, we mean a system for producing, acquiring, maintaining, distributing, and evaluating the astronomical data in computer-readable form. Although most of the computer-related operations will be done with the computing facilities available to the group, it has been emphasized that the resulting system should be machine-independent as much as possible, thus the compatibility being one of the most important goals in our work. Works currently under way are as follows.

    (a) Collection and validation of magnetic tapes containing astronomical data (mostly star catalogs).

(b) Production of the magnetic tape version for the catalog
    of stellar rotational velocities (Uesugi's revised
    catalog): In order to keep the input errors as few as
    possible, several validation schemes are being tried,
    including the use of interactive display terminals.
(c) Construction of an interactive data retrieval system:
    As the inter-university computer network is developed,
    automated data processing such as searching, sorting and
    extraction of data is expected to be a desirable
    procedure for the would-be users of the astronomical
    data system.
(d) Survey of the compatibilities and incompatibilities
    among the computer systems used by various institutions.
(e) Definitions of standards: We feel that the standards
    should be subject to an international agreement, and one
    is badly needed. However, if such an agreement is not
    expected to be reached soon, we must proceed with local
    standards which would be flexible enough for future
    changes. Currently, we are concerned about two types
    of standards. One is for the specification of the
    contents of the tape being distributed (whether the
    specification should be recorded within the tape itself
    or be written on a separate form). The other is for
    coding non-numerical informations such as the star name,
    the spectral type, etc..

5. ACKNOWLEDGEMENT

We are grateful to the organizing committees for making
various arrangements for our entry to the colloquium.

STELLAR DATA AND COMPUTING FACILITIES AT THE PULKOVO OBSERVATORY

D.D. Polojentsev

Pulkovo Observatory of the USSR Academy of Sciences

The Pulkovo Observatory computing laboratory began its activities in 1956. Originally it was equipped with card-punched machines. An essential work of processing astronomical data was carried out with the help of these machines ([1] and other).

With the development of the computing techniques in the USSR, the laboratory has been equipped with more modern computing devices. At present the Laboratory disposes of the third generation computer ES-1020 of the Ryad type, as well as the second generation computers Minsk-22 and Nairi-K.

The main data on the computers are given in the table below.

TABLE

Parameter	Computer Type		
	ES-1020	Minsk-22	Nairi-K
High-speed memory (op/sec)	20 000	15 000	2 000
Capacity of the memory (K)	64	32	16
Disk memory (N of devices)	2	–	–
Magnetic tape memory	4	8	–

C. Jaschek and G. A. Wilkins (eds.), Compilation, Critical Evaluation, and Distribution of Stellar Data. 195-197.
Copyright © 1977 by D. Reidel Publishing Company, Dordrecht-Holland. All Rights Reserved.

The computer ES-1020 is compatible with IBM computers at the level of the algorithmic languages, punch-cards and magnetic tape data. The computer Minsk-22 is compatible with other computers at the level of the languages ALGOL and FORTRAN and the punch-card data.

The Laboratory's staff consists of experienced astronomers who compile programs, skilled engineers and operators. Data is kept on punch-cards and magnetic tapes. Apart from original cata- logues and other data obtained from observations at Pulkovo and other observatories, the fundamental and general catalogues (FK4, N30, GC, PFKSZ, SAO, Yale, AGK3R etc) are available at the Labora- tory.

Direct data transmission in real time from the telescope to the computer is realized.

Astronomical ephemerides are necessary as a rule for reduction of astronomical data (e.g. observations). That is why basic parts of the "Annuaire Astronomique de l'URSS" were algorithmized and stored in the computers as a file of programs. The ephemeris time was chosen as the fundamental argument. While preparing for the work some formulas determining certain quantities were revised. For example, Besselian quantities C and D used to compute apparent places of stars were transformed to expressions of trigonometric series using fundamental arguments of the Lunar theory [2]. Wide application of modern computers in astronomy in our opinion re- quires a new form of the astronomical ephemerides. All the section of the astronomical ephemerides that are easily algorithmized and can be kept in the computer's memory might be significantly short- ened.

The astronomical ephemerides being the main source of most precise astronomical data must avoid detailed tables for hand cal- culations. The astronomical ephemerides are to be compiled taking into consideration the fact that the astronomer must be supplied with algorithms and corresponding test points.

The system of data centers which permits data to be stored in a machine-readable form is to be welcomed. In the USSR the Pul- kovo Observatory practically performs the function of the center for data on meridian astronomy and solar activity. Hence, great expenditure is saved for preparing data at every observatory.

The computing laboratory regularly exchanges data with the Astronomisches Rechen-Institut (Heidelberg) and the U.S. Naval Observatory and has contacts with the Stellar Data Center in Strasbourg.

The future development of the centers for astronomical data we see in cooperation in the field of data exchange and in centralization of data-keeping in certain national and world centers.

REFERENCES

1. M.S. Zverev, D.D. Polozhentsev.   Trudy Glavn. Astr. Obser.
   v. 72, 1958.
2. V.S. Gubanov.   Astron. J. USSR, v. 49, no. 5, 1972.

ASTRONOMICAL DATA FILES AT THE U. S. NAVAL OBSERVATORY: STAR
CATALOGUES, EPHEMERIDES, AND OBSERVATIONS

A. D. Fiala and P. K. Seidelmann

U. S. Naval Observatory

ABSTRACT

The history, contents, and method of distribution of the
astronomical data files of the Naval Observatory are described.
Some of the cooperative efforts to deal with the "data explosion"
are discussed.

INTRODUCTION

The Naval Observatory began using punched card equipment to
produce almanacs in 1940 and began accumulating astronomical data
in machine-readable form. The increased use of computers since
then by all astronomers has resulted in a great increase in the
generation and use of machine-readable data. Simultaneously
there has been a continuing program to transcribe older printed
observational catalogues into machine-readable form.

DATA AVAILABLE

The Naval Observatory has accumulated the following cate-
gories of data files in machine-readable form:

1. ephemerides of the Sun, Moon, planets, satellites, and
minor planets;
2. observations of same;
3. observations and catalogues of stars, both from the
Naval Observatory and other observatories;
4. precise time comparisons;

C. Jaschek and G. A. Wilkins (eds.), Compilation, Critical Evaluation, and Distribution of Stellar Data. 199-202.
Copyright © 1977 by D. Reidel Publishing Company, Dordrecht-Holland. All Rights Reserved.

5. catalogues of non-stellar or non-optical data;
6. tables and other information of astronomical interest.

These data files have been generated from

1. material produced as part of the regular publication of
almanacs;
2. basic material generated for producing the publications;
3. regular time service procedures;
4. regular observing (and reduction) programs;
5. transcription of older observations and catalogues from
printed form to machine-readable form;
6. observations or catalogues to be distributed for other
institutions;
7. staff research projects.

The data is stored in the form in which it was used, for
publication of research, etc. No effort has been made to use
standardized formats or to provide cross indexing or referencing.
This means that requests for data are responded to by copying all
the data of a file in the storage format for the period of time
covered by the file, and, if it is coordinate data, in the coordinate system in which it was prepared. There is no preparation
in a special format or extraction of specific data. For example,
coordinates of the Sun are available in apparent right ascension
and declination, or mean longitude and latitude, at daily interval,
1968-1980, or in geocentric spherical and rectangular coordinates
referred to 1950.0 at 4-, 10-, and 20-day intervals, 1800-2000.
The user would have to perform transformations to other coordinate
systems, or extractions for shorter time intervals, or interpola-
tions for other tabular intervals.

Almost all of the ephemerides and miscellaneous tables have
been published in the various publications of the Naval Observatory including annual almanacs, Circulars of the USNO, or the
Astronomical Papers of the American Ephemeris and Nautical
Almanac. Observations of stars at the USNO have been published
in Publications of the USNO, Second Series. Many of the star
catalogues are standard references available from many sources;
time data is circulated in Time Service Bulletins. Conversely,
most numerical data published by the USNO is available in
machine-readable form.

A list of data available, or in preparation, as of spring
1975 is published in USNO Circular No. 146, "Astronomical Data
in Machine-Readable Form." Requests for this Circular, or
inquiries about specific data not listed, may be directed to the
Superintendent, Naval Observatory, Washington, D. C. 20390.

INFORMATION EXCHANGES

The USNO provides information upon request, the terms being a three-for-one exchange of cards or tapes.

Some installations in the U. S. have data of interest to other users, but do not wish to be distributors either for lack of personnel or lack of appropriate computer capabilities. Then it is desirable for some other center to accept responsibility for distribution, with the stipulation that the original institution supply corrections and revisions as necessary. The USNO has done this in several instances, and is willing to consider more if the amount of material and the projected demand for it is within the capabilities of staff and computer time. However, this service may at a future time be provided by another federal agency, if the efforts of a recently appointed interagency task force are successful.

The proliferation of data and observations, especially by non-optical methods and space missions, raised interest over a decade ago in coordination of formatting and exchanging data. Initially, existing specialized data centers were supplemented through informal working groups composed of representatives from institutions having similar projects. Then there were formal and informal working groups within and between international organizations; later, formal operations centers were established, such as the Stellar Data Center at Strasbourg and the International Bureau on Astronomical Ephemerides at the Bureau des Longitudes, Paris. The latter is a reference and information center, rather than a data center, and thus it helps a potential user of data to contact the appropriate source directly and quickly, but does not provide data itself. Now, the U. S. Government has established an Interagency Coordinating Committee on Astronomy which has appointed a Task Force on Data. Part of the assignment of this task force is the investigation of establishing either data centers or information centers for U. S. Government agencies and possibly for non-government participants as well.

SUMMARY

The Naval Observatory continues to operate as a data center for astronomical data within the general limitations of its fields of expertise and the limitations of personnel and computer capabilities. Recognizing the problems of acquiring, storing, correcting, extending, editing, and distributing extensive astronomical data files, the Naval Observatory seeks to cooperate with all efforts to make knowledge of the data and the data itself available to astronomers around the world. To this end, it has

participated in several of the cooperative endeavors mentioned
previously. At the same time that the quantity of data available is growing, the observatory strives to improve the quality
and usefulness of the data which it has.

PART V

<u>EXISTING FACILITIES AND FUTURE ROLE OF DATA CENTRES</u>

SURVEY OF EXISTING FACILITIES

C. Jaschek

Observatoire, Centre de Données Stellaires – Strasbourg
(France)

The purpose of the present paper is to present a review of
the existing data centers.

If one knows nothing about a subject, one makes an enquiry,
and this is exactly what I have done. The list of addresses to
start with was provided by Dr. G.A. Wilkins, chairman of the
Working Group on Numerical Data of IAU Commission 5, who had
collected a list of data services. I want to thank all colleagues
who have answered the questionnaire, which is the basis of the
present paper; the answers (and the questionnaire itself) are
reproduced in Appendix A of this paper.

I shall start with the definition of "data service". The
data service should ensure that reliable data become available
promptly and conveniently to the user. Broadly one can specify
three different functions of the service, namely
>   a) compilation
>   b) evaluation or analysis
>   c) distribution

The data which are the subject of the work are values of various
parameters measured of one or various astronomical objects. Since
each datum, to be useful, must carry a source reference, data services are imbricated with documentation services.

It is easy to see that "data services" occupy a very definite
place in the organization of research.

We find in the first instance the colleague – or the team
of colleagues – who observes a number of astronomical objects and
describes the result of the observation by means of a series of

C. Jaschek and G. A. Wilkins (eds.), Compilation, Critical Evaluation, and Distribution of Stellar Data. 205-279.
Copyright © 1977 by D. Reidel Publishing Company, Dordrecht-Holland. All Rights Reserved.

TABLE 1

Data Dissemination Centers

HM Nautical Almanac Office Ephemerides - Herstmonceux (UK)

Nautical Almanac Office - Washington (USA)

National Space Science Data Center - Washington (USA)

Radio Observatory - Columbus (USA)

Centre de Données Stellaires - Strasbourg (France)

Institut de l'Astronomie Théorique - Leningrad (USSR)

Laboratory for Optical Astronomy - Greenbelt (USA)

Astronomisches Rechen-Institut - Heidelberg (GFR)

Bureau des Longitudes - Paris (France)

---

symbols for each object; for instance, numbers for magnitude, letters and roman numerals for spectral types. Each of these symbols represents a datum, and we will refer to these as "primary data", i.e. coming from observations.

A second instance is represented by a colleague - or a team of colleagues - who compiles the data regarding one type of objects (masses, radii and magnetic fields of Ap stars) or one type of data for different objects (UBV colors for all stars), or the documentation regarding one type of objects (a catalogue of planetary nebulae) or the documentation concerning one datum (stars observed photoelectrically). The compilation is usually accompanied by some evaluation or analysis of varying degrees of complexity, and the result is presented under the form of a "catalogue". This kind of activity will be described as "data evaluation" and the group which carries it out will be called a "data evaluation center" (DEC).

Usually, but not always, some kind of data distribution is carried out by the "data evaluation centers", the rule being a printed catalogue that crystallizes the knowledge at a given epoch.

A third instance is represented by an organization that combines the results from various "data evaluation centers". Its function lies mostly in the dissemination and the handling of the information through various more or less sophisticated techniques. It does not any longer handle primary data, i.e. original data, but mostly secondary data, i.e. data evaluated at one DEC. This is what we will call a data dissemination center, or more simply "data center" (DC).

The distinction between a DEC and a DC, which seems easy in theory, is somewhat difficult in practice. In Table 1, I have listed those institutes which seem to fall into the DC category.

The list of all DEC's is given in the Appendix. I have preferred to list all data services, even the smaller ones, because I think it is important to know what exists regardless of its volume. I would be grateful for any additions or corrections to this list.

Table 1 shows that the oldest data centers were the Institutes in charge of astronomical ephemerides; they also pioneered the use of modern computers in astronomy. Only recent years have seen DC's created on purpose.

Quite naturally the increasing volume of data to be handled as one proceeds from one instance to the next imposes an increasing complexity. Whereas DEC's tend to be handled by a single motivated individual on handwritten cards, the DC's are obliged to handle the

TABLE 2

Solar System

Earth, moon and artificial satellites
    Central Bureau of the International Polar Motion Service -
        Mizusawa (Japan)
    Department of Applied Mathematics and Computational Tech-
        niques - Moscow (USSR)
    Bureau International de l'Heure - Paris (France)
    Lunar Laser Data Management - Austin (USA)
    Lunar Data Center - Houston (USA)
Motions of planets and satellites
    HM Nautical Almanac Office Ephemerides - Herstmonceux (UK)
    Nautical Almanac Office - Washington (USA)
    Computing Laboratory of the Pulkovo Observatory (USSR)
    Astronomisches Rechen-Institut - Heidelberg (GFR)
    Bureau des Longitudes - Paris (France)
    Hydrographic Department of Japan - Tokyo (Japan)
Physical research, motions of smaller bodies, space exploration
    Planetary Research Center - Flagstaff (USA)
    IAU Planetary Photographs Center - Meudon (France)
    Minor Planet Center - Cincinnati (USA)
    The Minor Planets, Comets and Satellites Department -
        Leningrad (USSR)
    National Space Science Data Center - Washington (USA)

---

information by teamwork on large data handling facilities. This
can be shown very well in a tabulation of the different centers
made on the basis of the answers to the questionnaire. If one
classified the centers as DEC's and DC's, and the handling of the
information into two broad types, "handwritten" and "automatized",
one finds two facts, namely that half of the DEC's still process
handwritten information only and second, that no DC does so. It
is surprising that even today, when computing facilities are relatively common, half of the DEC's are still handling handwritten
information. The explanation lies probably in the ease with which
small amounts of information can be stored and handled.

Let us next review the situation in different parts of
astronomy. To do so I defined six areas, namely:
          Solar system (excluding the sun)
          Sun
          Stars
          Star systems and non-stellar objects
          Physics
          Bibliography
This division seems to be clear, although perhaps it is not entirely logical. I will next briefly examine each area, with
emphasis on stars, the main subject of the colloquium. Please
notice that a given center may be listed at more than one place.

SOLAR SYSTEM

We put into this group all centers dealing with solar system
bodies and interplanetary phenomena. We have listed in Table 2
the centers dealing with this subject. Those centers affiliated
directly with the ICSU Panel on World Data Centers are listed in
Table 3, together with the addresses. No questionnaires are provided for these centers in Appendix A.

TABLE 3

ICSU Panel on World Data Centers

---

1 - Comet tails, interplanetary scintillations and zodiacal light:
          Boulder, USA
          Kiev, USSR
          Munich, GFR

2 - Sporadic radio emissions from Jupiter:
          Boulder, USA
          Moscow, USSR

(continues)

TABLE 3, Continued

---

3 - Interplanetary magnetic fields:
                  Boulder, USA
                  Moscow, USSR
                  Tokyo, Japan

4 - Interplanetary electric fields:
                  Boulder, USA
                  Moscow, USSR

5 - Rotation of the earth:
                  Boulder, USA
                  Moscow, USSR

         Permanent services:
                  Bureau International de l'Heure
                  International Polar Motion Service

Addresses:

Boulder: World Data Center A for Solar Terrestrial Physics
           Environmental Data Service
           NOAA
           Boulder, Colorado 80302   USA

Kiev: Cathedra of Astronomy
           Observatornij per. 3
           KIEV 252053   USSR

Moscow: World Data Center B2
           Molodezhnaya 3
           Moscow 117296   USSR

Munich: World Data Center Cl
           Max-Planck-Institut für Astrophysik
           Föhringer Ring 6
           München 23, Federal Republic of Germany

Tokyo: Prof. T. Obayashi
           Institute of Space and Aeronautical Science
           University of Tokyo
           Komaba 4-6-1 Meguro-Ku
           Tokyo 153 Japan

Notes: ICSU = International Council of Scientific Unions.
           Boulder, USA is World Data Center A.
           Moscow, USSR is World Data Center B.
           The permanent services are listed separately; see Table 2.

TABLE 4

Sun

A – SOLAR AND INTERPLANETARY PHENOMENA

	Topic	WDC A	WDC B	Others
A.1	Sunspot Positions, Areas and Classification	Boulder, USA	Moscow, USSR	Freiburg, GFR
A.2	Sunspot Numbers	"	"	Zurich, Switzerland
A.3	Solar Magnetic Fields	"	Crimea	–
A.4	Hα Observations (other than flares)	"	Moscow, USSR	Rome, Italy Freiburg, GFR Meudon, France
A.5	Calcium Plages: Positions, Areas, Maximum Intensities	"	"	Arcetri, Italy Meudon, France
A.6	Combined and Special Optical Observations (Solar Maps, Prominences, Filaments)	"	"	–
A.7	Optical Observations of the Corona	"	"	Pic-du-Midi, France
A.8	Total Radio Flux Measurements	"	"	Toyokawa, Japan
A.9	Radio and Radar Maps of Solar Disk	"	"	–
A.10	Radio East-West Scans of Solar Disk	"	"	–
A.11	Solar X-ray and UV Background Levels	"	"	–
A.12	Energetic Solar Protons and Solar Electrons	"	"	–
A.13	Solar Wind	"	"	Tokyo, Japan
A.16	Monitoring of Total Solar Radiation	"	"	–

(continues)

TABLE 4, Continued

C - FLARE-ASSOCIATED EVENTS

Topic	WDC A	WDC B	Others
C.1 Hα Flares	Boulder, USA	Moscow, USSR	Meudon, France
C.2 Solar Magnetic Field in Active Regions and Their Short-Term Changes	"	"	-
C.3 Solar Radio Events, Fixed Frequency	"	"	Toyokawa, Japan
C.4 Solar Radio Spectrograms of Events	"	"	-
C.5 Solar X-ray Observations	"	"	-
C.6 Sudden Ionospheric Disturbances - Ground-Based Observations	"	"	Ondrejov, Czechoslovakia Slough, UK
C.7 Solar Protons and Electrons - Direct Measurement	"	"	-
C.8 Solar Protons - Riometer	"	"	Slough, UK
C.9 Solar Protons - Ionospheric Vertical Incidence Soundings	"	"	"
C.10 Solar Protons and Electrons - VHF Forward Scatter	"	"	"
C.11 Solar Protons - Other Types of Measurements	"	"	"
C.12 Solar, Ionospheric, or Aeronomical Rockets Launched During an Event	Greenbelt, USA	"	"
C.13 Cosmic Ray Ground Level Increases	Boulder, USA		Itabashi, Japan⎫ Umeå, Sweden ⎬

(continues)

TABLE 4, Continued

LIST OF ADDRESSES

Arcetri: World Data Centre C1
                    Osservatorio Astrofisico
                    Largo Fermi 5
                    50125 Firenze, Italy

Boulder: World Data Center A for Solar Terrestrial Physics
                    Environmental Data Service
                    NOAA
                    Boulder, Colorado 80302   USA

Crimea: WDC for Solar Magnetic Fields
                    may be reached via WDC B2 (Moscow)

Freiburg: World Data Centre C1
                    Professor Dr. K. O. Kiepenheuer
                    Schöneckstrasse 6
                    Fraunhofer Institut
                    D 78 Freiburg im Breisgau
                    Federal Republic of Germany

Greenbelt: World Data Center A for Rockets and Satellites
                    Code 601
                    Goddard Space Flight Center
                    Greenbelt, Maryland 20771   USA

Itabashi: World Data Centre C2
                    Dr. M. Wada
                    Cosmic Ray Laboratory
                    Institute of Physical and Chemical Research
                    Itabashi, Tokyo, 173, Japan

Meudon: World Data Centre C1
                    Dr. R. Michard
                    Observatoire de Meudon
                    92 Meudon, France

Moscow: World Data Center B2
                    Molodezhnaya 3
                    Moscow 117 296, USSR

Ondrejov: World Data Centre C1 for SID
                    Observator
                    Ondrejov-u-Prahy, Czechoslovakia

(continues)

TABLE 4, Continued

LIST OF ADDRESSES, Continued

Pic-du-Midi: World Data Centre C1
                          Observatoire du Pic-du-Midi
                          Service de la Couronne
                          65 Bagneres-de-Bigorre, France

Rome: World Data Centre C1
                          Prof. M. Cimino
                          Osservatorio Astronomico di Roma
                          Via del Parco Mellini 84
                          00136 Roma, Italia

Slough: World Data Centre C1
                          Appleton Laboratory
                          Ditton Park
                          Slough, Bucks., England SL3 9JX

Tokyo: Professor T. Obayashi
                          Institute of Space and Aeronautical Science
                          University of Tokyo
                          Komaba 4-6-1, Meguro-ku
                          Tokyo, 153, Japan

Toyokawa: World Data Centre C2 Solar Radio Emission
                          Professor H. Tanaka
                          Research Institute of Atmospherics
                          Nagoya University
                          Toyokawa, 442, Japan

Umeå: World Data Centre C1 for Cosmic Rays
                          Dr. S. Lindgren
                          Space Physics Group
                          University of Umeå
                          S-901 87 Umeå, Sweden

Zürich: World Data Centre C1
                          Prof. Dr. M. Waldmeier
                          Eidgenössische Sternwarte
                          Schmelzbergstrasse 25
                          8006 Zürich, Switzerland

    In Table 2 one sees several centers devoted to the study of the motions of the bodies of the solar system, whereas only two are oriented to physical aspects and both go mainly for planets. Apparently the natural satellites, other than the moon, receive

little attention. Curiously, meteors are even more neglected,
and an enquiry to the President of IAU Comm. 22 revealed the fact
that there exists no international meteorite center. [Note added
later on: At the IAU Assembly at Grenoble, Comm. 22 decided to
create a meteorite data center at Lund.]

The general impression is that the area is well covered.

SUN

This subject is almost entirely covered by the ICSU Panel on
World Data Centers for Solar–Terrestrial Physics (STP) and Rockets
and Satellites. Table 4 lists the different phenomena, the centers involved and their addresses. One can say that the coverage
is very satisfactory.

STARS

There exists a very great activity in this area, as shown by
the long list of centers given in Table 5. To this list one could
add further activities which have not been listed, as for example:
             - P. Kennedy - Mt. Stromlo (Australia) - MK spectral
                                                   types;
             - P. Bernacca - Asiago (Italy) - V sin i values;
             - G. Cayrel - Meudon (France) - abundances of metals in
                                               stellar atmospheres.

The situation seems thus to be very satisfactory, but a
closer look reveals some exceptions. I have shown previously
(Jaschek, 1968) that only in limited fields is the coverage of
the data up to date, meaning that a comprehensive catalogue has
appeared within the past two years, and this situation remains
unchanged. In Table 6, I have tried to summarize the situation
at mid–1976, quoting the date of the last comprehensive catalogue.

One special comment corresponds to the topic "spectral
peculiarities". The complexity of the problems is well illustra-
ted if one tries to answer the following question: where can one
obtain a list of subdwarfs (or of P–Cygni type objects, or of
metal-deficient stars)? (The only answer would be: ask Bidelman.)
Only for some special types of objects, like Be, Ap and Am stars,
etc. does a bibliography exist, and this lack of general coverage
is very unfortunate.

In passing, it would be convenient to estimate the speed
with which the information grows. It seems possible to assume
the growth to be of the exponential type, i.e.

TABLE 5

Stars

__Positions and motions__
    HM Nautical Almanac Office Ephemerides – Herstmonceux (UK)
    Nautical Almanac Office – Washington (USA)
    Astronomisches Rechen-Institut – Heidelberg (GFR)
    Institut de l'Astronomie Théorique – Leningrad (USSR)
    Computing Laboratory of the Pulkovo Observatory (USSR)
    Bureau des Longitudes – Paris (France)
    Space Science Center – Minneapolis (USA)
    Hydrographic Dept. of Japan – Tokyo (Japan)

__Radial velocities__
    Compilation of Supplement to RV Catalogue – Austin (USA)
    Bibliography of RV – Marseilles (France)

__Photometry__
    (Photométrie Stellaire) – Lausanne (Switzerland)
    Dearborn Observatory – Evanston (USA)

__Spectroscopic information__
    Astronomical Data File – Cleveland (USA)

__Double stars__
    Card Catalogue of Minima and Bibliography of Eclipsing
      Binaries – Cracow (Poland)
  * Centro de Dados de Estrelas Duplas – Rio de Janeiro (Brazil)
    Catalogues Complémentaires des Binaires Spectroscopiques –
      Toulouse (France)
    Card Catalogue of Eclipsing Binaries – Gainesville (USA)
    (Spectroscopic Binaries) – Victoria (Canada)
    (Visual Double Stars) – Washington (USA)
    (Cat. Orb. d'Etoiles Doubles Visuelles) – Brussels (Belgium)

__Variable stars__
    Variable Stars – Moscow (USSR)

__Miscellaneous__
    Department of Applied Mathematics and Computational
      Techniques – Moscow (USSR)
    (Pulsars) National Astronomy and Ionosphere Center – Ithaca
    (USA)

__Combined data__
  ** Centre de Données Stellaires – Strasbourg (France)
    National Space Science Data Center – Washington (USA)
    Laboratory for Optical Astronomy – Greenbelt (USA)

---

* This center has no publications yet.
** Copies of the Strasbourg files are also available through
   Potsdam, Zentralinstitut für Astrophysik (DDR).
Names in parentheses are given only for quick identification.

TABLE 6

Coverage of the Different Fields of Stellar Astronomy

Subject	Most Recent Catalogue
Parallaxes	
Trigonometric	Jenkins (1963)
Dynamic*	
Magnitudes	
Visual	Ochsenbein (1974), CDS
Magnitudes and colors	
Continuous updating	Observatoire de Lausanne, see also Centre de Données Stellaires
Spectrophotometry	
Scans	Breger (1976)
Spectral classification	
MK	Kennedy (1975), Centre de Données Stellaires
MK southern sky	Cowley and Houk (1975)
Spectral peculiarity	
	see text
Composition	
Metal abundance	Morel et al. (1976)
Detailed analysis*	
Position and proper motions	
	Haramundanis (1966)
Radial velocities	
	Abt and Biggs (1972)
Rotational velocity	
	Bernacca and Perinotto (1970-73)
	Uesugi and Fukuda (1970)
Radius**	
Binarity	
Visual	Jeffers, Van den Bos and Greeby (1963)
Δm in visual binaries	Wierzbinski (1969)
Orbits of binaries	
Visual	Finsen and Worley (1970)
Spectroscopic	Batten (1967)
Eclipsing	Koch, Plavec, Wood (1970)
Variability	
Main elements	Kukarkin et al. (1969), continuously updated
Bibliography*	
Magnetic fields	
	Babcock (1958)

* The subject was covered some time ago and coverage is outdated.
** No coverage was ever attempted.

TABLE 7

Growth Index α

Datum	Objects	(Epoch)	Objects	(Epoch)	α
Radial velocities	15 000	(1953)	25 000	(1972)	0.0269
Spectral types	15 400	(1964)	22 500	(1975)	0.0180
Photometry UBV	22 000	(1968)	25 000	(1972)	0.0320

$$N = N_0 \, e^{\alpha(t - t_0)}$$

where N_0 and t_0 are constants and α is the growth index, giving
the relative increase per annum. Please notice that this includes
only the growth of the number of new objects being measured. The
figures of Table 7 show that for rapidly advancing fields the
annual rate is of the order of 3%. If one were to measure the
total growth, i.e. including re-measurements of already known objects, one would arrive at about 8%. This figure comes from a
still unpublished study and shows that every year a large number of objects is being re-measured. Although this is sometimes
necessary, it implies also that much duplication exists, which is
mostly due to the ignorance of the results published since the
last printed catalogue. One say squarely that everything which
is not contained in the last catalogue is lost for the astronomer
because of "bibliographic inaccessibility". A partial solution
is found in the "Bibliographie Cayrel" which the Centre de Données
Stellaires is beginning to publish; the bibliography permits one
to know if a given star has been mentioned in the recent literature. I say purposely that this constitutes a palliative, because
one has to look star by star, since very often one also does not
know which objects belong to the class one is searching for.

A glance at Table 6 shows that we are still very far from
having a continuous updating, and this is probably the most
pressing need to fill in the years to come.

STAR SYSTEMS AND NON-STELLAR OBJECTS

We have grouped here all work done on clusters and galaxies,
and also on non-stellar objects because of the small number of
centers dealing with such topics. Table 8 quotes the relevant
centers.

The first impression one gets is that the effort made to
synthetize our knowledge of non-stellar objects (which is admit-
tedly a very wide field) is definitely much less organized than

TABLE 8

Star Systems and Non-Stellar Objects

Clusters
   Open clusters - Moscow (USSR)
   Globular clusters - Moscow (USSR)
   Globular clusters - Ontario (Canada)

X-ray objects
   Laboratory for High Energy Astrophysics - Greenbelt (USA)
   High Energy Astrophysics Division - Cambridge (USA)

Galaxies
   Extragalactic Astronomy Group - Austin (USA)

Non-stellar objects
   Radio Observatory, Ohio State - Columbus (USA)
   Variable stars,- Moscow (USSR)

---

the work on stars. Since I have no firsthand knowledge of the field, I offer only a few comments, hoping that the audience will provide the answers.

1) For non-stellar objects exists the compendium of Dixon, which provides the background for a very large number of objects. The main question is if this is enough, and what remains to be done. It would seem that a number of things - above all bibliographic references - would be very welcome. Some work on special non-stellar objects has been started in France (see the paper by Lortet in this volume), but is as yet unpublished.

2) Despite the very encouraging start given by the Westerbork Group for radio astronomy, it seems also that the field is largely open for somebody trying to cope with the rapidly increasing information.

PHYSICS

Although the centers occupied with the compilation of physical data really fall outside the scope of this listing, I have listed in Table 9 some centers which are well known to many colleagues because of their active role in our field. It should be stressed that many more centers do exist. A complete listing of them can be found in the "International Compendium of Numerical Data Projects", CODATA (1969).

TABLE 9

Physics

JILA Atomic Collision Cross Section Information Center - Boulder
    (USA)
Data Center on Atomic Transition Probabilities - Washington (USA)
Data Center on Atomic Line Shapes and Shifts - Washington (USA)
Atomic Energy Level Data Center - Washington (USA)

BIBLIOGRAPHY

Under this general title I have listed the different information services which exist in Astronomy. They are included for
completeness since they are not primarily data centers. The list
is given in Table 10. We find in the first place the abstract
services; nothing needs to be said about them, since they are working very well, providing a regular and much appreciated coverage
of the literature.

There comes next a group of services which are of a very specific nature, namely the:

a) Bureau International d'Information, which supplies the announcements of catalogues published and where to get them.

b) The undertaking of Davis, who prepares a list of programs for
   carrying out analytical computations on computers.

c) IAU Depositories for variable stars, which keep original unpublished data.

d) INSPEC, which provides a bibliography and an index of astronomical catalogues.

A third group finally concerns those enterprises which provide
bibliography for isolated objects (mostly stars), which are Bidelman's data file and the Cayrel bibliography. The first was being
punched by Parsons, whereas the second will be published soon by
the Centre de Données Stellaires (see the paper by Ochsenbein and
Spite in this volume). Both enterprises are a very great step
forward in the rational use of the existing information. The only
warning I must issue is that they do not constitute an automatic
substitution of the old selected bibliographies (like Merrill and
Burwell's bibliography of Be stars), unless the stars are coded in
a system which permits the user to find the members of the group
he is interested in.

TABLE 10

Bibliography

Abstracting services
    Astronomy and Astrophysics Abstract – Heidelberg (GFR)
   * Referatni Journal – Moscow (USSR)
    Bulletin Signalétique – Paris (France)

Information services and miscellanea
    Bureau International d'Information sur les Ephemerides
        Astronomiques – Paris (France)
    (Computer Programs) – Chapel Hill (USA)
    IAU Comm. 27 Depositories for Photoelectric Observations of
        Variable Stars – London (UK)
    International Information Services for the Physics and
        Engineering Communities – London (UK)

Specialized bibliographies
    Astronomical Data File – Cleveland (USA)
    (Bibliographie Stellaire) – Meudon (France)
    Lunar Data Center – Houston (USA)

---

* No questionnaire for this service can be provided. The address
is: Referatni Journal – Baltijskaya ul., 14   125219 Moscow
A-219   USSR.
Names in parentheses are given only for quick identification.

This completes our survey of existing data centers. I think
that they cover satisfactorily large portions of our science – the
nearest objects being logically the best covered, and the most
distant galaxies the least well covered. We could all be very
satisfied with this state of affairs, were it not for the disturbing fact that few of our colleagues use the information stocked
at these centers. Astronomy in a certain sense is still a science
with a nineteenth-century mentality, so that novelties are only
slowly accepted. It seems to me that a long effort is due to
show to all colleagues the usefulness of data center services.

REFERENCES

Abt H.A. and Biggs E.S. (1972) Bibliography of stellar radial
    velocities. Latham Process Corp., New York.
Babcock H.W. (1958) Astroph. J. Suppl. $\underline{3}$, 141.
Batten A.H. (1967) Publ. Dominion Astroph. Obs. $\underline{13}$, 119.
Bernacca P.L. and Perinotto M. (1970-73) Contr. Oss. Padova in
    Asiago nos. 239, 250 and 294.
Breger M. (1976) In press.

CODATA (1969) International compendium of numerical data projects.
    Springer Verlag, Berlin, Heidelberg, New York.
Cowley A.P. and Houk N. (1975) University of Michigan catalogue
    of two-dimensional spectral types for the HD stars. Dept. of
    Astronomy, University of Michigan, Ann Arbor. Vol. I.
Finsen W.S. and Worley C.E. (1970) Circ. Republic Obs. Johannes-
    burg 7, 203.
Haramundanis K. (1966) Star catalog. Smithsonian Astrophysical
    Observatory, Smithsonian Institution, Washington, D.C.
Jaschek C. (1968) Publ. Astron. Soc. Pacific 80, 654.
Jeffers H.M., van den Bos W.H. and Greeby F.M. (1963) Index
    catalogue of visual double stars. Publ. Lick Obs. 21.
Jenkins L. (1963) General catalogue of trigonometric stellar
    parallaxes. New Haven, Conn., Yale University Observatory.
Kennedy P. (1975) Information bulletin of the Stellar Data
    Center, Strasbourg, no. 10.
Koch R.H., Plavec M. and Wood F.B. (1970) Publ. Univ. Pennsylvania,
    Astr. Ser. XI.
Kukarkin B.V., Kholopov P.N., Efremov Yu.N., Kukarkina N.P.,
    Kurochkin N.E., Medvedeva G.I., Perova N.B., Fedorovich V.P.
    and Frolov M.S. (1969) General catalog of variable stars.
    Astron. Council of Acad. Sci. USSR, Moscow.
Morel M., Bentolila C., Cayrel G. and Hauck B. (1976) In press.
Ochsenbein F. (1974) Astron. Astroph., Suppl. Ser. 15, 215.
Uesugi A. and Fukuda I. (1970) Contr. Inst. Astroph. Kwasan Obs.
    Kyoto no. 189.
Wierzbinski S. (1969) Contr. Wroclaw Obs. no. 16.

APPENDIX A

The following questionnaire was sent to all data centers.

1. Name of the Data Center (DC), group or section.

2. Affiliation of the DC.

3. Address.

4. Name of the Director, or person responsible.

5. Main objective of the DC.

6. Publications issued by the DC.

7. Publication (if existing) in which the activities of the DC are described.

8. Does your center store the information on hand written cards, punched cards, magnetic tapes and/or others?

9. Are your data files available through interchange, loan, sale?

10. Additional information.

The answers received are reproduced in what follows. Centers are arranged by country, and countries are listed in alphabetical order.

The author would be grateful if omissions and/or corrections were brought to his attention, since he intends to update the file regularly.

1. Département d'Astrométrie et de Méchanique Céleste.

2. Observatoire Royal de Belgique.

3. avenue Circulaire, 3 - B-1180 BRUXELLES   BELGIQUE.

4. Dr. J. DOMMANGET.

5. - Catalog of visual binary orbits (including all orbits,
      even the most ancient ones, for each binary). This
      documentation is prepared for different statistics.
    - Catalog of ephemerides of radial velocities (see point 6).
      The purpose is to provide radial velocity observers with
      all necessary informations for systematic observing
      programs for visual doubles, with the ultimate aim of
      determining the position of the ascending node (researches
      on the orientation of orbital planes), the masses and paral-
      laxes of the system, etc.

6. Catalogue d'éphéméride des vitesses radiales relatives des
    composantes des étoiles doubles visuelles dont l'orbite est
    connue.
    Communication de l'Observatoire Royal de Belgique, série
    B, N° 15, 1967.
    Second catalogue d'éphémérides des vitesses radiales relatives
    des composantes des étoiles doubles visuelles dont l'orbite
    est connue (manuscript in preparation).

7. -

8. Catalog of orbits: on handwritten cards; the most recent
    orbits are on punch cards.
    Catalog of ephemerides of radial velocities: graphs plotted
    on cards.

9. Copies of Comm. O.R. ser. B, No. 15 are available upon request.
    Information on ancient orbits can be provided upon request.

10. Updating of the catalogs is done by the staff.

1. Centro de dados de estrelas duplas.

2. Observatório Nacional.

3. Rua General Bruce,   586 - São Cristóvão,   20.000 Rio de
    Janeiro,   BRASIL.

4. Dr.   Ronaldo Rogério de Freitas Mourão.

5. Collection of all data of visual double stars: astrometric,
    photometric, spectroscopic and orbital; parallaxes and
    masses.

6.

7. No.

8. Punched cards, magnetic tapes.

9.

10. The future role of the Double Stars Data Centre is the anal-
    ysis and informations of the data of visual double stars.

1. Globular Clusters.

2. David Dunlap Observatory (University of Toronto).

3. Richmond Hill, Ontario, Canada L4C 4Y6.

4. Dr. Helen Sawyer Hogg.

5. A card index of all references on globular clusters and a
    separate index of all references to variables in globular
    clusters.

6. Catalogue of Variables in Globular Clusters (The Third
    Catalogue) DDO Publications 6 no. 3 (1973).
    Bibliography on Globular Clusters, First Supplement, DDO
    Pub. II no. 12 (1963).

7. None.

8. Hand written cards.

9. I will supply as much information as I can on specific
    clusters or subjects. It would not be feasible to supply
    entire files.

10. Dr. Christine Coutts Clement works on variables in globular
    clusters here and has access to my files also, so she could
    supply information in my absence.

1. None.

2. Dominion Astrophysical Observatory.

3. 5071 W. Saanich Rd., Victoria B.C., V8X 3X3, Canada.

4. A. H. Batten.

5. To provide an up-to-date, comprehensive, and critical
    catalogue of the orbital elements determined for spectro-
    scopic binaries at such intervals as appear warranted by
    the activity in the field.

6. Catalogue of S.B. orbital elements issued in Publications of
    Dominion Astrophysical Observatory 13, 119 (1967).

7. -

8. Hand-written cards, but conversion to tapes or punched cards
    under active consideration.

9. No, but we are ready to give information from them at any
    time.

10. Computer available at Observatory. Two people work on
    project. Cordial informal relations maintained with
    Pedoussaut's group in Toulouse.

1. Observatoire de Marseille.

2. -

3. 2, Place Le Verrier - Marseille - France.

4. Dr. M. Barbier.

5. Bibliography of stellar radial velocities published after 1970.

6. -

7. Bulletin CDS (Strasbourg) $\underline{8}$, 12 (1975).

8. On punched cards.

9. Through the "Stellar Data Center" at Strasbourg.

10. -

1. Bibliographical Star Index.

2. Observatoire de Paris, Section de Meudon.

3. Observatoire, 92190 MEUDON - FRANCE.

4. R. Cayrel.

5. Supply bibliographic file for stellar objects. See article
    in information bulletin No. 10 of the CDS, Strasbourg. Also
    this volume, paper number 27.
6. None.

7. Information Bulletin of the "Centre de Données Stellaires",
    No. 10.

8. On notebooks.

9. Sale through the "Centre de Données Stellaires".

10. -

1. IAU Planetary Photographs Center (IAUPPC).

2. Observatoire de Paris - 92190 - MEUDON FRANCE.

3. Same.

4. A. Dollfus, R. Servajean.

5. Collecting the photographs of planets collected by tele-
    scopic observations throughout the world and the planetary
    and lunar imaging documents resulting from the space
    missions.

6. Circulars.

7. Booklet, "The IAU planetary photographs centers."
    Transactions of the International Astronomical Union since
    1961.

8. True pictures in negative or positive or both, and punched
    cards.

9. For consultation at the center for scientific work, or for
    reproduction for scientists, free of charge if for small
    amount. Special case if large amount.

10. 1 curator, 1 documentalist, 1 photographer, 1 technician.

1. Bureau des Longitudes.

2. -

3. 77 Avenue Denfert-Rochereau, 75014 PARIS - France.

4. B. Morando.

5. Publication of ephemerides.

6. La Connaissance des Temps (once a year).
    Ephémérides Nautiques (once a year)
    Annuaire du Bureau des Longitudes (once a year).

7. -

8. Cards or 9-track tapes.

9. Available free of charge if on cards. If on tape send a
    blank tape.

10. -

1. Bureau International d'Informations sur les Ephémérides
    Astronomiques.

2. Bureau des Longitudes.

3. 77 Avenue Denfert Rochereau   75014 - PARIS - FRANCE.

4. Dr. B. Morando.

5. The purpose of the Information Bureau is to provide informa-
    tion to the international scientific community on the avail-
    ability of astronomical ephemerides, catalogues of star posi-
    tions and lists of reduced positional observations for use
    in astronomical and space research. The Bureau:
    - receives details of astronomical ephemerides that are
      currently available or in course of preparation;
    - maintains indexed lists of such ephemerides and of
      cooperating institutions and individuals;
    - publishes from time to time information cards giving summary
      lists, details of significant new ephemerides and other
      relevant information; and
    - answers requests for information about the availability of
      ephemerides in printed or machine-readable form.
    In addition the Bureau may, in appropriate circumstances:
    1) make recommendations to cooperating institutions as to
       convenient standards for the specification and supply of
       ephemerides; and
    2) receive details of, and answer questions about, new
       determinations of astronomical constants, but the Bureau
       is not expected to carry out original work in this field.

6. Information cards no. 1 to 125 between 1971 and 1976.

7. -

8. The center does not store data but information on printed
    cards on the data available.

9. The information collected by the Bureau is distributed free
    of charge to the institutions that request it. It is given
    on information cards, sequentially numbered.

10. -

1. Bureau International de l'Heure.

2. Federation of Astronomical and Geophysical Services.

3. 61, avenue de l'Observatoire - Paris 75014 - France.

4. Bernard GUINOT.

5. 1 - The BIH evaluates and makes available the International
       Atomic Time and the Universal Time Coordinated, which
       are the worldwide basis for timing the events, in
       accordance with the recommendations of the Conférence
       Générale des Poids et Mesures.
    2 - The BIH evaluates the coordinates of the instantaneous
       terrestrial pole and the universal time UT1, from all
       the relevant data (classical astronomy, measures on
       artificial satellites, and, in the near future, lunar
       distances).

6. Monthly Circular D } giving the results.
    BIH Annual Report }

7. None.

8. On punched cards (recent data). Files on magnetic disks.

9. Magnetic tapes containing the observational data can be sent
    (subject to reimbursement of the expenditures).

10. -

1. Centre de Documentation Scientifique et Technique du CNRS.

2. Centre National de la Recherche Scientifique.

3. 26 rue Boyer, 75971 Paris Cedex 20.   France.

4. J. H. d'Olier, Director.

5. The aim of the CNRS Documentation Centre is to provide
   research scientists as well as specialists, academics,
   doctors, engineers, industrialists, students, etc., in France
   and abroad, with the scientific and technical documentation
   necessary for their research and activities. Founded in 1939
   to run a scientific library and produce an abstracts journal,
   the Centre has continued to increase and develop its activities.
   Thanks to a staff of 400 technicians, engineers, doctors and
   field specialists, it plays a pilot role in the elaboration
   and application of documentary techniques. The Centre, with
   its world-wide multidisciplinary coverage, processes documents
   in the fields of Exact Sciences, Life Sciences, Earth Sciences
   and Technology. It has a growing data file drawn from a
   variety of scientific publications. More particularly, the
   Bulletin Signalétique Section 120 file, which includes
   Astronomy, was automated in 1972. It now (end 1976) contains
   about 90 000 items with a yearly increase of about 18 000
   items.

6. Bulletin Signalétique du CNRS (see PASCAL. Plan de classement)
   51 Sections, in particular Section 120. Astronomie. Physique
   spatiale. Géophysique. (ten issues per year).

7. See "Informascience".

8. Magnetic tapes density 800 or 1600 BPI; tracks, 9; code,
   EBCDIC; label, not present or present (OS standard).

9. Sale.

10. Overall personnel: 400 technicians, engineers, doctors.

   Participation, in company of major foreign documentation
   centres, in the work of those organisations whose aim is to
   coordinate questions of scientific documentation: UNISIST,
   ISO, ICSU-AB (in particular the development of an international
   classification scheme for Physics within the frame-work of the
   Working Group in Physics), FID, EUSIDIC, CIDST.

   Conversational under study.

1. Centre de Données Stellaires.

2. Observatoire de Strasbourg.

3. 11, rue de l'Université - 67000 Strasbourg - France.

4. Dr. C. Jaschek.

5. To collect the most important stellar data; to analyze them critically; to make them available to the scientific community.

6. Information Bulletins, Nos. 1-11, two issues per year, 1971-76.

7. See Information Bulletins.

8. Preferably 9 track, 1600 BPI magnetic tapes.

9. Data are available on request, with small service charges. See Information Bulletins for details.

10. -

1. Catalogues complementaires des binaires spectroscopiques.

2. Observatoires du Pic du Midi et de Toulouse.

3. Observatoire de Toulouse, 1 Avenue Camille Flammarion, 31500 Toulouse - France.

4. Dr. A. Pedoussaut.

5. To inform specialists during the long intervals between successive general catalogs. Thus between the Moore-Neubauer (1948) catalog and the Batten (1967) catalog, we published 9 Complementary Catalogs. Since the publication of Batten's (1967) catalog, we have published 3 complements, and a fourth one will be published in the near future. The Observatoire de Toulouse will have published a total of 13 catalogs between 1952 and 1976.

6. - Annales de l'Observatoire de Toulouse, T. XXI (1952) to T. XXXII (1968).
   - Astr. Astrophys. Supple. $\underline{4}$ (1971); $\underline{10}$ (1973); to appear (1976).

7. Some very short remarks in Astr. Astrophys. $\underline{4}$ (1971).

8. On handwritten and on punched cards.

9. We have always answered all requests and will continue to do so.

10. Three persons (N. Ginestet, J.M. Carquillat, R. Nadal) have contributed to the assembling of the Complementary catalogs. When the latest complement is ready, we think it can be integrated with that of the "Centre de Données Stellaires" at Strasbourg.

1. Astronomy and Astrophysics Abstracts (AAA).

2. Astronomisches Rechen-Institut.

3. Astronomisches Rechen-Institut, Moenchhofstr. 12-14,
    6900 Heidelberg, Fed. Rep. Germany.

4. Dr. L.D. Schmadel - Editor-in-chief.

5. AAA is devoted to the recording, summarizing and indexing of
    astronomical publications throughout the world. It is pre-
    pared under the auspices of the International Astronomical
    Union (according to a resolution adopted at the 14th General
    Assembly in 1970). AAA aims to present a comprehensive
    documentation of literature in all fields of astronomy and
    astrophysics. Two volumes with 14000-15000 abstracts are
    scheduled to appear per year. While each volume contains an
    author index and a subject index, the magnetic tapes con-
    taining the index informations will be used to produce separ-
    ate index volumes at intervals of a few years.

6. AAA Vol. 1 - 13 (1969-1975)   (two volumes annually).

7. Reviews of single volumes in different astronomical journals.

8. The text of the abstracts are written with IBM 72 Composer.
    Only the index informations are stored on magnetic tapes.

9. The printed volumes may be obtained from Springer-Verlag,
    1000 Berlin.

10. a) Sorting programs are written in PL/I. The computations
       are carried out on an IBM 370/168.

    b) 14 persons: astronomers, translator for the Russian
       language, typists, key punching operator (some of them
       only for part-time job).

    c) AAA is a member of the ICSU Abstracting Board.

1. (Data center for astrometric catalogues).

2. Astronomisches Rechen-Institut Heidelberg.

3. Mönchhofstr. 12-14 - 6900 Heidelberg 1 - Fed. Rep. Germany.

4. T. Lederle.

5. At the Astronomisches Rechen-Institut, the improvement of
    the FK4 and its extension to fainter stars, resulting in a
    new fundamental catalogue (FK5) is being carried out. In
    connection with this project, a data file of astrometric cat-
    alogues - observing, compiled, and fundamental - in machine-
    readable form is necessary. Some observatories have con-
    tributed by providing catalogues or observational data on
    magnetic tapes; the greatest part of the file which consists
    of more than 300 catalogues at present have been key-punched
    at the Institute. Information on the catalogues which are
    available is given in the Bulletin du Centre de Données
    Stellaires and in the B.I.I.E.A. Information Cards.

6. -

7. Bulletin du Centre de Données Stellaires No. 1, p. 10, 1971.

8. On punched cards (80-column) and magnetic tapes (9 tracks,
    EBCD code).

9. See B.I.I.E.A. Information Card No. 78, 1973.

10. Access to IBM 360/44 and 370/168.

1.   Central Bureau of the International Polar Motion Service.

2.   International Latitude Observatory of Mizusawa.

3.   Mizusawa-shi, Iwate-ken, 023 Japan.

4.   Dr. Shigeru YUMI.

5.   Main objectives of the IPMS are the continuous determination
     of the precise position of the Earth's rotation axis which
     is required for geodetic and astronomical purposes, that is,
     for determining the positions of the observers and those of
     the stars observed, and the promotion of studies on relevant
     problems of the pole motion as the results of the continuous
     determination mentioned above, that is, a study of the sys-
     tem of astronomical constants, particularly the constants of
     nutation and aberration, positions of the stars, as well as
     a geophysical study of a constitution of the Earth and a
     possible continental drift.

     To meet with the objectives of the IPMS, the Central Bureau
     collects worldwidely all the observational data, determines
     pole motion, analyses it and distributes the data and results.
     Results obtained by the new techniques like the satellite
     Doppler observation, laser ranging and so on, are also dealt
     with.

6.   Annual Report of the IPMS for the years 1962, . . . , 1973.
     (Once a year).
     Monthly Notes of the IPMS 1962-Nos. 1-12, . . . , 1975-Nos.
     1-11. (12 issues a year).

7.   The Europa Year Book 1976 : A World Survey, London.
     The World of Learning 1957-76, London.
     Federation of Astronomical and Geophysical Services (FAGS)
     (A booklet describing the activities of FAGS on the occasion
     of its tenth anniversary). UNESCO, 1966.

8.   Original data of observations are recorded on hand or type
     written sheets. All of them are also punched on cards in a
     machine readable form. They are also recorded on magnetic
     tapes. Photocopies of original data on a fiche film will be
     available in the near future.
     Specifications on magnetic tapes are :

(continued)

(continued)

	at present	in the near future
Number of tracks	7	9
Recording	BCD(transformable to the Code IBM 26)	Code IBM 29 and EBCDIC
Density	800 or 556 bits/inch	800 or 1,600 bits/inch
Block length	variable	variable
Block interval	19mm	--

9. a) Data files are available through cost sale.
    b) Annual Report and Monthly Notes of the IPMS in a printed
       version are available free of charge by duly request at
       present. Selling system in the future time is under
       examination.

10. a) Computer
       TOSBAC 3400   Model 51

    b) Personnel
       Computer   :   7 person
       Center business   :   7 person

    c) Links with
          B I H

1. Astronomical Division.

2. Hydrographic Department of Japan.

3. Tsukiji-5, Chuo-ku, Tokyo 104, Japan.

4. Akira M. Sinzi.

5. Improvement of ephemerides and catalogues.

6. Data Report of Hydrographic Observations
    Series of Astronomy and Geodesy (issued annually).

7. None.

8. Magnetic tape, 9 tracks, 800 or 1600 BPI.

9. Mostly loan.

10. -

1. Card Catalogue of minima and bibliography of eclipsing
   binaries.

2. Astronomical Observatory of the Jagiellonian University.

3. ul. Kopernika 27, 31-501 Krakow, POLAND.

4. Prof. dr Andrzej Zięba.

5. The collecting of data for the preparation of ephemerides of
   eclipsing binaries.

6. "Rocznik Astronomiczny Obserwatorium Krakowskiego",
   International Supplement /SAC/ founded by T. Banachiewicz,
   one issue per year.

7. -

8. Hand written cards.

9. The data are available through interchange.

10.

1. Institut d'Astronomie de l'Université de Lausanne et
2. Observatoire de Genève.

3. 1290 Sauverny – SWITZERLAND.

4. Prof. B. Hauck.

5. Collection and homogenisation of photometric data.

6. Rapports internes; rapports internes spéciaux.

7. See papers on the Centre de Données Stellaires, Strasbourg.

8. Magnetic tapes (7 channels, 9 from 1976 on, BCD).

9. Through the Centre de Données Stellaires, Strasbourg.

10. Eight people are engaged in this work. Links are with
    Strasbourg and IAU Comm. 25 and 45.

1. H.M. Nautical Almanac Office.

2. Royal Greenwich Observatory.

3. Herstmonceux Castle, Hailsham, East Sussex BN27 1RP, UK.

4. Dr. G.A. Wilkins (Superintendent).

5. The principal data activities of H.M. Nautical Almanac Office
    are currently as follows:
    1. The preparation for publication of almanacs and tables
    containing astronomical data for use by astronomers, naviga-
    tors, surveyors and others.
    2. The computation of ephemerides and other astronomical
    data for particular places or projects; such data are used
    for research purposes, or are published in newspapers, diaries,
    etc., or are used as evidence in legal cases.
    3. The prediction, reduction and analysis of occultations of
    stars by the Moon in order to study the motion of the Moon, the
    rotation of the Earth, and related matters; most of the obser-
    vations are made by amateur astronomers in many different
    countries. Predictions for occultations by the Moon of plan-
    ets, radio sources and X-ray sources, etc., as well as of
    occultations of stars by planets and satellites are also made.
    4. The maintenance of a library of astronomical ephemerides,
    star catalogues, etc., on magnetic tape for use by the staff
    of the Office and of other departments of the Royal Greenwich
    Observatory. Copies may be made available to others under
    appropriate conditions.

6. Annually: Astronomical Ephemeris, Nautical Almanac, Air
    Almanac, Star Almanac for Land Surveyors. From time to
    time: navigational and mathematical tables.

7. Activities are reported in Transactions IAU in report of
    Commission 4 (Ephemerides).

8. Mainly magnetic tape (most 1/2-inch, 7-track, 556 rows/inch).

9. Exchange, loan or sale depending on the circumstances, but
    the Office may decline to meet requests.

10. The Office has access to the ICL 1903T computer system of
     the Royal Greenwich Observatory and cooperates very closely
     with the Nautical Almanac Office of U.S. Naval Observatory.

1. Depositories for photoelectric observations of variable stars.

2. IAU Comm. 27.

3. See under 10.

4. Dr. W. S. Fitch – Steward Observatory – The Univ. of Arizona – Tucson, Ari 85721 – USA.

5. See title.

6. Circular Letters. The Depositories contain 43 series with more than 85,000 observations.

7. Trans. IAU 15-A, 313 (1973).

8. Printed or handwritten material.

9. Yes. Copies available at cost, on request.

10. Copies of files can be obtained from either

Dr. E.W. Maddison, Librarian   Dr. V.P. Tsessevich
Royal Astronomical Society   Astronomical Observatory
Burlington House   Shevchenko Park
London, WIV ONL, England   Odessa GSP 714, USSR

1. INSPEC (International Information Services for the Physics and Engineering Communities).

2. The Institution of Electrical Engineers.

3. Savoy Place, LONDON WC2R OBL, England.

4. D. Barlow, Director.

5. To provide a comprehensive English language abstracting service in physics (including astronomy and astrophysics), electrical and electronic engineering, computers and control.

6. The services offered by INSPEC are as follows:
   Physics Abstracts (including approx. 6000 abstracts/year in astronomy and astrophysics), Current Papers in Physics, Electrical and Electronics Abstracts, Current Papers in Electrical and Electronics Engineering, Computer and Control Abstracts, Current Papers on Computers and Control, Cumulative Indexes, Key Abstracts, SDI (Selective Dissemination of Information), Topics, Magnetic Tapes, and On-Line Retrieval (through Lockheed DIALOG, ESA/RECON, and CAN/OLE).

7. INSPEC Matters available from INSPEC.

8. Information is stored on magnetic tapes. The distribution format for INSPEC tape services is based on ISO 2790, the international standard format for bibliographic data interchange. For more details contact INSPEC Retrieval Services Department.

9. Magnetic tape copies of the INSPEC data base are for sale.

10. Bibliography and Index of Astronomical Catalogues, 1951-75, will be available in magnetic tape form and may be regularly updated. For further details contact Mike Collins, F.R.A.S. (Senior Information Scientist), INSPEC, Station House, Nightingale Road, HITCHIN, Hertfordshire SG5 1RJ, England.

1. David S. Evans.

2. Personal scientific enterprise.

3. Department of Astronomy, University of Texas, Austin,
    Texas 78712, USA.

4. David S. Evans.

5. Compilation of supplement to RV Catalogue.

6. None: proof copies circulated.

7. None.

8. Typescript.

9. Xeroxes available.

10.

1. Extragalactic Astronomy Group.

2. Dept. of Astronomy, The Univ. of Texas at Austin.

3. Univ. of Texas, R.L.M. 16-316, Austin, TX 78712, USA.

4. Prof. G. de Vaucouleurs.

5. Compilation and reduction of standard systems of morphologic, diameter, magnitude, colors, 21 cm and continuum flux, and radial velocity data on bright galaxies.

6. Reference Catalogue of Bright Galaxies (UT Press 1964).
    New Reference Catalogue of Bright Galaxies (UT Press 1976).

7. None.

8. All 3 forms are used (mag. tape specs on request).

9. Prior to 1976 mag. tape copies were provided free of charge. After 1976 mag. tape copies will be distributed at cost (possibly through UT Press).

10. Core of group is G. de V., A. de V. and H. G. Corwin, Jr. (with intermittent NSF or UT funding). Variable number of graduate students are employed as Research Assistants depending on NSF funding (lapsed at end 1975). Working relations for radio data exist with Nançay, NRAO, and other centers.

1. Lunar laser data management.

2. Department of Astronomy, University of Texas at Austin.

3. Austin, Texas 78712, U.S.A.

4. Dr. Peter J. Shelus.

5. The overall objective of the lunar laser data management pro-
    ject is to transform the raw observations produced by the LURE
    team station into useful observations of the round trip time
    delay from these stations to specific points on the lunar sur-
    face. This general process is constituted by a complex series
    of individual tasks. The Observatory records consist of laser
    firing times, photon detection times, system calibration and
    environmental information. The first step consists of data
    identification. Since the majority of detected photons are
    ambient light rather than real signal, this is necessarily a
    statistical process, the details of which are given by Abbot,
    Shelus, Mulholland and Silverberg (Astron. J. 78, 784, 1973).
    This process produces the individual measured time delays, as
    well as an estimate of the associated uncertainty. These in-
    dividual measures are then compressed into normal points for
    use in discussions of the observations. It is the responsi-
    bility of this project to provide these data monthly to LURE
    Team members or subsequent analysis, and to deposit them on a
    NASA approved semi-annual schedule to the archives of the
    National Space Science Data Center.

6. No formal publication series.

7. Astron. J. 78, 784 (1973); Astron. J. 80, 154, (1975);
    Astron. J. 80, 723, (1975).

8. The project maintains its own copies of the raw photon detec-
    tions on magnetic tape, the filtered observations on magnetic
    tape and punched cards, and the normal points on magnetic tape
    and punched cards.

9. Since these data are openly available from the NASA National
    Space Science Data Center, Goddard Space Flight Center, Code
    601, Greenbelt, Maryland 20771, USA (non-US requestors contact
    COSPAR World Data Center A at the same address), the lunar
    laser data management project attempts to avoid direct dis-
    tribution of these data wherever feasible. Under special cir-
    cumstances, we have made direct transmission of the normal
    point data on either punched cards or magnetic tape without
    charge, so long as such distribution is consonant with the
    NASA approved public data release schedule.

(continued)

(continued)

10. Magnetic tapes at the University of Texas at Austin are
    written with a Control Data Corporation 6600 computer. The
    project has, however, developed software capability to write
    seven track tapes that are compatible with IBM/360 BCD con-
    figuration. This activity is funded by NASA as a part of
    the lunar laser ranging program.

1. JILA Atomic Collision Cross Section Information Center.

2. Joint Institute for Laboratory Astrophysics.

3. JILA A 305
    University of Colorado
    Boulder, CO 80309
    USA

4. Dr. Earl C. Beaty.

5. The Joint Institute for Laboratory Astrophysics (JILA)
    maintains an Information Center for collecting information
    on low energy atomic collisions. The areas we are now
    actively working in are: electron collision cross sections;
    photo absorption cross sections; electron swarm and rate
    data; ion-molecule reaction rates. These data are of par-
    ticular importance to applications in astrophysics, atmos-
    pheric physics and plasma physics. Our products consist of
    bibliographies, data compilations and critical reviews.

    All Information Center reports will be announced to those on
    the Information Center's mailing list and will be made
    available by publication
    1) in a recognized professional journal;
    2) as a report in the National Standard Reference Data
       Series (these reports will be for sale by the Super-
       intendent of Documents);
    3) or in the JILA Information Center Reports Series
       (these will usually be distributed free upon request).

    The choice of the mechanism for circulation of such informa-
    tion will depend on the nature of the information, the
    expected users, and the availability of the publication to
    the users. We will try then to insure the maximum distribu-
    tion of this material to the most likely users, with the least
    inconvenience to them.

6. Unscheduled.

7. None recently.

8. Methods of storing data are chosen to suit the problem.

9. Data is distributed to qualified users on request.

10.

1. High Energy Astrophysics Division.

2. Smithsonian Astrophysical Observatory; Smithsonian Institution.

3. 60 Garden Street, Cambridge, Massachusetts 02138, U.S.A.

4. Prof. R. Giacconi.

5. We plan to maintain a data bank of the observations obtained
    by the UHURU X-ray observatory covering the period from
    December 1970 to April 1973. The purpose of this archive is
    to allow astronomers access to these early X-ray satellite
    observations when they can add significantly to future
    analysis or discoveries.

6. Important results are published in the Center for Astrophysics
    reprint series.

7. None.

8. Data are stored on magnetic tapes and important summaries on
    microfilm and printout.

9. Data files are not available through interchange, loan, or
    sale. For access see questions 5 and 10.

10. Access to the UHURU observations can be obtained through the
    Principal Investigator, Prof. R. Giacconi, under the guide-
    lines of the Guest Investigator program.

1. (None)

2. Morehead Observatory.

3. Dept. of Physics and Astronomy, University of North
    Carolina, Phillips Hall, Chapel Hill, N.C. 27514, U.S.A.

4. Prof. M.S. Davis.

5. Preparation of a list of programs (<u>not data</u>) for carrying
    out analytical computations on computers.

6. None.

7. None.

8. Lists are being compiled and will be available in printed
    form.

9. They will be available to all interested persons or groups.

10. Large, regional, teleprocessing Computation Center available.
    Central computer: IBM 165. Local computers connected to 165:
    IBM 360/75, 370/155II.

1. Minor Planet Center.

2. Cincinnati Observatory, University of Cincinnati.

3. Observatory Place, Cincinnati Ohio 45208, USA.

4. Paul Herget.

5. Collect all minor planet observations. Publish Minor Planet
    Circulars. Compute elements and ephemerides.

6. Minor Planet Circulars (No. 3938 since 1947).

7.

8. Magnetic tapes. We have more than 160,000 obsv. since 1939
    in card-image.

9. Each request is considered upon its merits.

10. The Cincinnati Observatory is also interested in determining
    new plate constants for the Astrographic Catalogue. The
    Bordeaux zone has been completed.

1. Astronomical Data File.

2. Warner and Swasey Observatory of Case Western Reserve
    University.

3. 1975 Taylor Rd., E. Cleveland, Ohio 44112, U.S.A.

4. William P. Bidelman.

5. The objective of the data center is to facilitate access to
    all significant known data concerning non-solar system ob-
    jects. In many cases actual data are included in the data
    file, and in all cases it is intended to give references to
    all important published work. Some 60,000 objects are
    presently contained in the data file. Even in its currently
    very incomplete form the file has proved extraordinarily
    useful, and it is believed to be the most comprehensive such
    file in existence.

6. None issued by the center as such. However two general
    catalogues have been issued by Dr. Stephenson of the Warner
    and Swasey Observatory which should be mentioned: "A General
    Catalogue of Cool Carbon Stars," W. and S. Publ. 1, No. 4, and
    "A General Catalogue of S Stars," W. and S. Publ. 2, No. 2.

7. Trans. IAU 13A, 1020 (1967).
    Publ. Leander McCormick Obs. 16, p. 172-186 (1971).
    IAU Symp. No. 50, p. 288 (1973).

8. At present the file is on hand-written cards. However, it
    has been partially punched at the U. of Texas (see Bull.
    AAS 6, 217, 1974).

9. I am happy to attempt to answer inquiries concerning any non-
    solar system astronomical object. The data files are open to
    all visitors.

10. At the moment the data-file work is supported only by the
    Observatory, with only one part-time assistant.

1. Radio Observatory.

2. Ohio State University.

3. 2015 Neil Avenue, Columbus, Ohio 43210, USA.

4. Dr. Robert S. Dixon.

5. The objectives of our group are:
    a) To make astronomical data available in computer-readable
        form.
    b) To organize astronomical information and data so it can
        be simply comprehended, not only by astronomers, but by
        all scientists and engineers who have need for it.
    c) To avoid duplication of effort in data gathering and
        program writing.
    d) To distribute widely useful materials and data to anyone
        who needs them. These materials include catalogs, maps,
        charts, photographs, books and overlays.
    e) To promote the use of computers in science.
    We maintain and distribute these two main publications:
        The Master List of Radio Sources, containing discovery
    (and rediscovery) data for all the 70,000 known radio sources.
        The Master List of Non-Stellar Objects, containing
    discovery (and rediscovery) data for all the 185,000 known
    non-stellar objects.
        In addition, we distribute other small catalogs, maps,
    charts, photographs and books, as described in our catalog
    of publications.

6. Catalog of Publications; new editions of which are issued
    periodically.

7. None.

8. Punched cards, magnetic tapes and computer printouts. All 7
    and 9 track tape formats available on the IBM 370 computer
    system can be used.

9. All data files are available for sale, at a cost which equals
    our costs of duplication and distribution. We are also will-
    ing to exchange data files with others who have data that we
    need.

10. There are 5 people engaged in data organization here, and we
    have access to IBM 370/168, 1130 and 1620 computers. There
    are no formal links to other data groups, but we maintain
    informal periodic contact with other interested groups and
    individuals.

1. Dearborn Observatory.

2. Northwestern University.

3. 2131 Sheridan Road, Evanston, IL 60201, USA.

4. Prof. William Buscombe.

5. Stellar Spectral Classifications.

6. MK Spectral Classifications, 1974.

7. -

8. IBM cards; 7-track tapes, 556 bpi.

9. Sale.

10. Vogelback Computing Center, CDC 6400.

1. Planetary Research Center.

2. Lowell Observatory.

3. P.O. Box 1269, Flagstaff, Arizona 86001, USA.

4. Dr. William A. Baum.

5. The broad objective of the Planetary Research Center is to
    advance the state of knowledge concerning the bodies in the
    solar system, particularly as it relates to their origin and
    evolution.
        Specific current objectives include:
        Systematic investigation of visible atmospheric and
    surface phenomena on Mars, Jupiter, and Venus through the
    analysis of planetary photographs.
        Studies of planets, satellites, and other solar system
    objects through photometric and polarimetric observations,
    including those that augment image analysis.
        Intensive photographic surveillance of Mars, Jupiter, and
    Venus by means of a worldwide network of ground-based
    stations.
        Collection, organization, and cataloguing of planetary
    photographs from all sources.
        Maintenance of a facility for guest investigators and
    the supplying of photographs and data to scientists at other
    institutions.
        Cooperation with or participation in space missions for
    which our experience is relevant.
        Advancement of instrument technology in areas contribu-
    ting to the other objectives above.

6. (a) Approximately 100 papers published since 1965 in scien-
    tific journals; list available.
    (b) Status Reports prepared twice a year; copies on request.

7. Trans. IAU 15A, 199-200 (1973). IAU Symposium 65, pp. 241-251
    (1974). Planetary and Space Science 21, 1511-1519 (1973).
    Icarus 12, 435 (1970).

8. Catalog of 120,000 planetary photographs is on punched cards.
    Printouts (12 vols. to date) available. Magnetic tapes planned.

9. Printouts are easily available. Tapes will eventually be.
    Punched cards can be duplicated with greater effort.

10. Qualified guest investigators have access to the photographic
    collection, catalogs, image measuring instruments, computer,
    and other facilities. Application to the director should
    include a brief statement of the scientific purpose.

1. Card Catalogue of Eclipsing Binaries.

2. Department of Physics and Astronomy, University of Florida.

3. Gainesville, Florida 32611, U.S.A.

4. Frank Bradshaw Wood.

5. Literature References on Eclipsing Binaries.

6. A Finding List for Observers of Eclipsing Variables.
    A catalogue of graded photometric studies of close binaries.
    Published at irregular intervals and infrequently (4 issues
    of former and one of latter to date).

7. Brief discussion in Volume 1, Vistas in Astronomy, Ed. A.
    Beer (1975).

8. Hand written cards (although the Finding List has been put
    on punched cards or tape at several institutions).

9. Copies of the cards for any specific systems are supplied
    on request.

10.

1. Laboratory for High Energy Astrophysics, X-ray Astronomy
    Group.

2. NASA/Goddard Space Flight Center.

3. Code 661, Greenbelt, Maryland 20771, USA.

4. John Arens and Richard Rothschild.

5. To collect in one place a listing of all X-ray observations
    of galactic sources in the energy range .1 to 500 keV.
    Included is source name, position, detector energy range,
    spectral resolution, temporal resolution, observation date,
    detector area, observation length, payload vehicle,
    reference, and comments.

6. Goddard X-document once per year.

7. None yet.

8. Data stored on tape and punched cards.

9. Possible, but no specifics have ever been generated.

10. -

1. Laboratory for Optical Astronomy.

2. NASA, Goddard Space Flight Center.

3. Code 671, Greenbelt, MD 20771, USA.

4. Dr. Jaylee Mead.

5. To establish a computerized astronomical data retrieval system.

6. No regular publications.

7. B.A.A.S. 6, 21 (1974); B.A.A.S. 8, 346 (1976).

8. Preferably 9-track, 1600 BPI magnetic tapes.

9. Interchange or loan.

10. Access to GSFC IBM 360-75 + 91 computers; two persons engaged in project; work closely with National Space Science Data Center.

1. National Space Science Data Center.

2. NASA - Goddard Space Flight Center.

3. Code 601, Greenbelt, Maryland 20771, USA.

4. Dr. James Vette.

5. To further use of reduced data and to serve as an active
    repository for such data, especially in space science.

6. Data Catalog of Satellite Experiments, NSSDC 71-20.
        "   "   "   "   "   , Supplement No. 1
    to NSSDC 71-20, NSSDC 73-11.

7. -

8. Any tape specification or punched cards can be handled.

9. Single copies supplied at no cost to scientific workers.

10. Work closely with NASA - Goddard Space Flight Center's
    Laboratory for Optical Astronomy.

1. Lunar Data Center.

2. Lunar Science Institute.

3. 3303 NASA Road No. 1, Houston, TX 77058, U.S.A.

4. Dr. Robert O. Pepin, Director of the Institute; Mrs. Frances
    B. Waranius, Librarian; Mrs. Mary Ann Hager, Photo/Map
    Librarian.

5. To assemble, catalog, and generally make accessible to the
    scientific community as complete a collection of lunar data
    as possible.

6. Lunar Science Information Bulletin No. 1-8 (1974-1975) appr-
    oximately 4 times per year.

7. Geoscience Information Society Proceedings 5, 71-75 (1975).
    Special Libraries 66, 407-410 (1975).

8. The information is stored in a variety of published formats.
    Books, documents, photos, maps, journal articles, etc.

9. Most materials in the Data Center are available on loan to
    qualified investigators either by individual requests or
    through standard inter-library loan procedures.

10. We maintain and operate a Moon Literature Bibliography which
    is stored on a computer disk. Search capabilities exist for
    authors or title key-words. Searches are done on request.
    At present, there is no charge for the search service.

1.

2. Cornell University - "National Astronomy and Ionosphere Center".

3. Cornell University, Space Sciences Bld., Ithaca NY 14853, USA.

4. Dr. Y. Terzian.

5. Compilation of observational data on Pulsars.

6. Astrophysics and Space Sciences (1976).

7.

8. Computer cards, magnetic tapes, paper printout.

9. Available on request.

10.

1. Space Science Center.

2. Univ. of Minnesota.

3. Minneapolis, MN 55455 - USA.

4. W. J. Luyten.

5. Proper Motions of faint stars.

6. Proper motion survey with the 48-inch Schmidt telescope
    I-XLIII (to date).

7. No. I of above.

8. Handwritten cards, magnetic tape, computer print-outs.

9. Available to anyone who wishes to come here.

10. All salient data will be made available to NASA Data Center,
    Greenbelt, MD 20771, USA.

1. Astrometry and Astrophysics Division.

2. U.S. Naval Observatory.

3. U.S. Naval Observatory, Washington, D.C. 20390, USA.

4. C.E. Worley.

5. Collection, correction, and dissemination of visual double star data to astronomers throughout the world.

6.

7. Bulletin A.A.S. <u>6</u>, 217, (1974).

8. Punched cards and magnetic tapes (tape can be 7 or 9-track, from 556 to 3200 B.P.I.).

9. Interchange (user sends blank tape and receives data in return).

10. We have access to an IBM 360/45. Two people are engaged in the work.

1. Atomic Energy Levels Data Center.

2. National Bureau of Standards.

3. Washington, D.C. 20234, U.S.A.

4. W. C. Martin.

5. The main objective is the compilation and publication of critically evaluated data on atomic energy levels and spectra. The compiled data include energy levels (with J and g values, configuration and term description), wavelengths and classifications of spectral lines. The bibliographic files of references also include papers with other data such as hyperfine structure, isotope shifts, stark effect, etc. A compilation of the energy levels of iron in all 26 ionization stages was published recently, and similar work for chromium is in progress. A compilation of the levels of the rare-earth elements is nearing completion. Charlotte E. Moore is continuing her <u>Selected Tables of Atomic Spectra</u> (energy levels and multiplet tables), having recently completed the tables for neutral oxygen. Publication of a <u>Bibliography on Atomic Energy Levels and Spectra (July 1971 through June 1975)</u> is planned during 1976.

6. Compilations and bibliographies on atomic spectral data.

7. CODATA Bulletin <u>14</u>, 112-115 (1975).

8. Spectroscopic data on punched cards and magnetic tapes (7 track BCD mode, 800 bpi, even parity). Bibliographic files on punched paper tapes (Dura typewriter) and magnetic tapes.

9. We will answer specific inquiries. Address to Dr. Lucy Hagan.

10. We have access to a UNIVAC 1108 Computer. Three to four persons work in the Center.

1. Data Center on Atomic Line Shapes and Shifts.

2. National Bureau of Standards.

3. National Bureau of Standards, Rm. A267, Bldg. 221, Washington,
    D.C. 20234, U.S.A.

4. Dr. W. L. Wiese.

5. The two main objectives of the center are: (1) the collection
    and cataloging of all literature relevant to the broadening
    and shift of atomic spectral lines; and (2) the preparation
    and publishing of bibliographies and critical reviews of
    various topics in atomic line broadening.

6. Bibliography on Atomic Line Shapes and Shifts:
        NBS Special Publication 366 (1972)
        NBS Special Publication 366,Supplement 1, (1974)
        NBS Special Publication 366, Supplement 2, (1975)
    Critical reviews:
        J. Phys. Chem. Ref. Data (in press) (experimental Stark
            broadening parameters for non-hydrogenic spectral lines
            of neutral atoms)
        J. Phys. Chem. Ref. Data (in press) (experimental Stark
            broadening parameters for non-hydrogenic spectral lines
            of ionized atoms).

7. Bulletin American Astronomical Society, No. 1, first issue
    each year, 1970-1975 (for example Vol. 4, No. 1, p. 137-139
    (1972).

8. Bibliographic information stored on punched cards and
    magnetic tape (for internal use only).

9. Data and references available upon request by phoning (301)
    921-3374 or by writing to the address given above. Bib-
    liographies may be purchased from U.S. Government Printing
    Office, Washington, D.C. 20402. Reprints of J. Phys. Chem.
    Ref. Data publication may be purchased from Subscription
    Service Department, American Chemical Society, 1155
    Sixteenth Street, NW, Washington, D.C. 20056.

10. Four people working in data center (1 full time, 3 part time).

1. Data Center on Atomic Transition Probabilities.

2. National Bureau of Standards.

3. National Bureau of Standards, Rm. A267, Bldg. 221,
   Washington, D.C. 20234, U.S.A.

4. Dr. W.L. Wiese.

5. To collect and catalog all relevant literature, to extract
   and analyze the numerical data, and to prepare and publish
   bibliographies and tables of "best" values.

6. Bibliography on Atomic Transition Probabilities:
       NBS Special Publication 320 (1970); Supplement 1 (1971);
       Supplement 2 (1973).
   Critical data compilations:
       NSRDS-NBS 4, Vol. I (1966) (hydrogen through neon)
       NSRDS-NBS 22, Vol. II (1969) (sodium through calcium)
       NBS Technical Note 474 (1969) and Atomic Data $\underline{1}$, 1-17
           (1969) (Ba I, II)
       J. Phys. Chem. Ref. Data $\underline{2}$, 85-120 (1973) (forbidden
           lines of scandium through nickel)
       J. Phys. Chem. Ref. Data $\underline{4}$, 263-352 (1975) (scandium
           and titanium)
   Critical analyses of systematic trends in atomic oscillator
   strengths:
       Astrophys. J. Suppl. $\underline{23}$, No. 196, 103 (1971) (helium
           through magnesium isoelectronic sequences)
       Phys. Rev. A, Feb. 1976 (in press) (spectral series of
           lithium isoelectronic sequence); data tables to be
           published in J. Phys. Chem. Ref. Data.

7. Bulletin American Astronomical Society, No. 1, first issue
   each year, 1970-1975 (for example Vol. 4, $\underline{\text{No. 1}}$, p. 137-139
   (1972).

8. Information is not automated at present.

9. Data and references available by phoning (301)921-3374 or by
   writing to us. Bibliographies and NBS compilations may be
   purchased from U.S. Government Printing Office, Washington,
   D.C. 20402. Reprints of J. Phys. Chem. Ref. Data publications
   may be purchased from: Subscription Service Department, American Chemical Society, 1155 Sixteenth Street, NW, Washington,
   D.C. 20056.

10. Four people working in data center (1 full time, 3 part time).

1. Nautical Almanac Office.

2. U.S. Naval Observatory.

3. Nautical Almanac Office, U.S. Naval Observatory, Washington, D.C. 20390, USA.

4. Dr. P. K. Seidelmann.

5. The data center was established to provide data upon request in maching readable form, primarily in America. The center specializes in ephemerides and astrometric star catalogs, which are the fields of specialization of the personnel of the office. Additionally, the center does serve as a distribution method for other astronomers who do not wish, or have the facilities, to provide copies of their machine readable data.

6. U.S. Naval Observatory Circular No. 146 (a new circular is issued when it appears necessary).

7. Annual Reports of U.S. Naval Observatory in Bulletin of the American Astronomical Society.

8. Data is stored and provided on punched cards or magnetic tapes (either 7 track, 556 or 800 bpi, or 9 track, 800 or 1600 bpi).

9. The data is available on an exchange basis, three new tapes, or cards, for each requested tape, or card.

10. This data center was established in conjunction and cooperation with Her Majesty's Nautical Almanac Office and the Astronomisches Rechen - Institut.

1. Computing Laboratory of the Pulkovo Observatory.

2. Main Observatory of the U.S.S.R. Academy of Sciences (Pulkovo).

3. Pulkovo Observatory, 196140 Leningrad M-140, Pulkovo, U.S.S.R.

4. Dr. D.D. Polozhentsev.

5. Processing of the astronomical data, mostly in the field
    of meridian astronomy and solar activity.

6. None.

7. Paper for IAU Colloquium No. 35 (1976).

8. We store data on punched cards and magnetic tapes (9 tracks,
    800 bits per inch), on the main results of astrometric
    observ. and star catalogues.

9. Through interchange.

10. See the paper for IAU Colloquium No. 35.

1. Institut de l'Astronomie théorique de l'Acad. Sci. URSS.

2.

3. 10, quai de Koutouzoff – Leningrad 192187 – USSR.

4. V. A. Broumberg, V. K. Abalakine, V. A. Shor.

5.

6. Astronomitcheskiĭ Ejegodnik SSSR; Efemeridy malykh planet – every year.
   Algoritmy Nebesnoĭ Mekhaniki – 6-8 issues per year.

7. –

8. Hand written cards, punched cards.

9. Interchange, loan.

10.

1. The Minor Planets, Comets and Satellites Department.

2. The Institute of the Theoretical Astronomy of the USSR
    Academy of Sciences.

3. Institute of the Theoretical Astronomy, Naberezhnaja
    Kutuzova 10, 192187 Leningrad, USSR.

4. Dr. V.A. Shor.

5. The Department of Minor Planets, Comets and Satellites is
    engaged in the following kinds of activities: a) the obser-
    vations of minor planets and comets using 16" double astro-
    graph of the Crimean Astrophysical Observatory; b) the
    improvement of the orbital elements, the computation and
    publication of opposition ephemerides of all catalogued
    (numbered) planets; c) the researches on spatial and dynam-
    ical structure of the asteroid belt; d) the improvement of
    star positions from minor planet observations; e) the com-
    putations of the definitive orbits of comets, the investiga-
    tions of the evolution of comet orbits caused by planet per-
    turbations and nongravitational effects; f) the studies of
    the motion of some natural satellites.

    The Ephemeris volume annually issued by the Department
    since 1947 contains: a) orbital elements of all numbered
    planets, their mean opposition magnitudes $B(a,0)$ and abso-
    lute magnitudes $B(1,0)$; b) dates of the oppositions;
    c) opposition ephemerides comprising 70 days; d) extended
    ephemerides of bright planets $(B(a,0) \leqslant 11^m5)$ covering 200
    days; e) extended ephemerides of sume unusual planets having
    some specific features such as approaches to the Earth, the
    great eccentricity, inclination; f) the information on the
    insufficiently observed planets.

6. The Ephemerides of Minor Planets, published annually.

7. The activities of the Center are systematically reflected
    in the triennial reports of Commission 20 in the Transactions
    of the IAU.

8. The information is stored on punched cards and magnetic tapes
    in the binary-decimal code used for BESM-4 electronic computer.

9. The data files may be available through interchange.

10. The Ephemerides of Minor Planets are prepared in collabora-
    tion with the Minor Planet Center (Cincinnati Observatory)
    and the Astronomical Observatory of Latvian State University.

1. Astronomical Council of the USSR Academy of Sciences.

2. Sternberg State Astronomical Institute of Moscow University.

3. 109017, Moscow 17, ul. Pyatnitskaya 48, USSR.

4. Prof. Dr. B.V. Kukarkin.

5. The work on the collection, storage, analysis and distribu-
    tion of data on variable cosmic objects is carried out by the
    Astronomical Council of the USSR Academy of Sciences, to-
    gether with the Sternberg State Astronomical Institute of
    Moscow University, since 1946 on behalf of the International
    Astronomical Union. Here is a brief account of this work.
    1. Compilation of a full set of cards on published variable
    star studies as well as those sent to us in the form of
    letters and preprints. During recent years the work was
    extended by inclusion of pulsars, X-ray sources, non-stable
    quasars and galactic nuclei. Our work is not restricted to
    collecting information; it is mainly devoted to thorough
    critical analysis of all data. Most important data are
    included in the Catalogue only after careful evaluation of
    all the information. These data are systematically put on
    punched cards and magnetic tapes.
    2. The General Catalogue of Variable Stars is usually pub-
    lished once a decade (the latest, third, edition was pub-
    lished in 1969-71).
    3. During the intervals between the publication of the Gen-
    eral Catalogue, several Supplements to it are published. For
    instance the Third Supplement to the latest edition will be
    published by summer 1976.
    4. Besides this special reference books are published from
    time to time: catalogues of suspected variables, lists of
    variable stars arranged in the order of right ascensions etc.
    5. Each year or two, we compile name-lists of variable stars
    recently discovered or investigated. Twenty lists were pub-
    lished since 1946.
    6. By now (1976) the cards contain data on more than 40,000
    objects.
    7. The General Catalogue and Supplements contain the follow-
    ing data: equatorial and galactic co-ordinates, references
    to the main study of the star and to the paper containing the
    finding chart, type of variability, limits of variations,
    elements of variations, spectral types. Numerous remarks
    contain more special data.
    8. The work on improving the classification of variable
    stars and other non-stable celestial objects is continuously
    under way.

(continued)

6. The General Catalogue of Variable Stars, 3rd edition 1969-71,
    three volumes. The Supplements to the Catalogue (three
    issues 1971-76). The bulletin "Variable Stars" (13 Numbers
    1971-75 and 9 Supplements 1971-75).

7. See Sky and Telescope 40, 355 (1970).

8. Detailed information about all variable stars on hand
    written cards. The information selected for the Catalogue,
    on punched cards and magnetic tapes.

9. In 1976. Not ready yet.

10. Three persons work all the time. Eight persons take part in
    the work from time to time as experts, if necessary.

1. Astronomical Council of the USSR Academy of Sciences.

2. Sternberg State Astronomical Institute of Moscow University.

3. 109017, Moscow 17, ul. Pyatnitskaya 48,   USSR.

4. Prof. Dr. B.V. Kukarkin.

5. Since 1959 the work on the collection and analysis of data
   on globular clusters of our Galaxy and (during recent years)
   of the Magellanic Clouds and the Andromeda Nebula (M31) is
   carried out at the Astronomical Council of the USSR Academy
   of Sciences and the Sternberg State Astronomical Institute
   of Moscow University. The main objective of our work is to
   undertake a careful analysis of all the information and to
   put diverse and heterogeneous data on a unique system. In
   1974 the book "Globular Star Clusters" and the catalogue of
   129 globular clusters were published. The second enlarged
   edition of the book and the catalogue is in preparation (it
   is planned to be in print in 1978). We collect permanently
   information based on the account of all available new data.
   The cards contain the following information: equatorial and
   galactic co-ordinates, integrated V-magnitudes, colour
   indices B-V, U-B and V-I, colour excesses E(B-V), information
   on intermediate- and narrow-band photometry, various parameters of colour-magnitude diagrams, spectral classes, metal
   abundances, numbers of red giant and horizontal branch stars,
   numerical characteristics of stellar concentration in the
   clusters, tidal and core radii, flattening, variable stars
   (numbers according to types), properties of variable stars
   (periods etc.), presence of anomalous stars, X-ray sources
   etc., radial velocities and proper motions, apparent and
   true distance moduli, data on the brightest stars of the
   giant branch, integrated absolute V-magnitudes, number of HI
   atoms in the direction of globular clusters, and additional
   information.
        Apart from the data on 129 globular clusters which we
   consider to be undoubtedly globular, we also analysed data
   on some clusters which are not proven to be globular.

6. The book "Globular Star Clusters", Moscow 1974. "General
   Catalogue of Globular Clusters of Our Galaxy,"Moscow 1974.

7. -

8. On hand written cards.

9. We are ready to send information in written form.

(continued)

(continued)

10. Two persons work all the time. Six persons take part in
     the analysis of the information. We constantly have links
     to many persons, observatories and institutions.

1. Department of Applied Mathematics and Computational Techniques.

2. Astronomical Council of the Academy of Sciences of the USSR.

3. Moscow 109017, Piatnitskaya, 48, USSR.

4. Prof. A.G. Massevitch.

5. The main objectives and used data files:
   1. Automatic reduction of satellite frames. The whole complex of programs for obtaining positions and moments of observation for satellites is based on the SAO star catalogue stored on magnetic tape. Our variant of the catalogue contains α_{1950}, δ_{1950}, μ_α, μ_δ, M_v, Sp for each star with number N SAO. An economical code packing method is used.

   Now this catalogue and the complex are adjusted for ES-computer system.
   A system of storing in files on magnetic tape of satellite data is prepared.
   2. Ephemeris service for satellite observing stations.
   3. Astrophysical problems:
      - A complex of programs in the field of stellar structure and stellar evolution
      - A complex of programs for analysis of the spectra of magnetic stars
      - Stellar statistics programs.

6. Nautchnye informatsyi N 30, 34 (nonperiodically).

7. -

8. punched cards ГОСТ - 6198,
   magnetic tapes 35 mm width.

9. Our data on punched cards are available through interchange, as are printed catalogues of α, δ, τ for satellites.

10. We have a computer M-222 (M-20 type):
    rate - $3 \cdot 10^4$ op/sec., main storage - 16 kwords (one word - 45 digits). We have an access to a BESM-6 computer too: rate - 10^6 op/sec., main storage - 32 kwords.

1. Department of Astrophysics.

2. Astronomical Council of the USSR Academy of Sciences.

3. 109017, Moscow 17, Pyatnitskaya 48, USSR.

4. Dr. O.B. Dluzhnevskaya.

5.   Open clusters

     1. Work on the analysis of published data on open
clusters has started at the Astronomical Council of the USSR
Academy of Science about three years ago. From the published
data for about a thousand open clusters were chosen only
those having UBV-photometric data for all cluster members.
The membership of stars in those clusters was redetermined
by both UBV-photometry data and proper motions or radial
velocities in order to obtain homogeneous data allowing
thorough statistical investigations of the ages and evolutionary status of the cluster and its integral parameters.

     2. Several computer programs have been evaluated using
results of stellar models computations for massive stars by
B. Paczynski [1], I. Iben [2] and V. Varshavsky and A.
Tutukov [3] ($16 \geqslant M/M_o \geqslant 0.8$) at different evolutionary
stages. Those programs allow estimation of masses and ages
of individual cluster members together with the errors of
these estimations and also the corresponding probabilities
for each pair of values (if for certain stars the theory
allows several values). Some statistical characteristics,
like the luminosity function, the mass function, integral
colours, integral luminosity and others have been determined
for each cluster.

     3. Now our catalog includes data for 65 open clusters
(about 7500 cluster members) and contains the following
information:
     For each cluster:
NGC or IC number
References
color excess
distance modulus
remarks
H-R-diagram in the $M_v - (B-V)_o$ - plane
H-R-diagram in the $\lg Te - \lg L/L_o$ - plane
number of member stars
mass function
luminosity function

distribution of members with age
mean age
integrated color
integrated magnitude (B-V)
integrated magnitude V
integrated mass from all member stars

For each star of the cluster:
number
color indices: V, B-V
        (Te - effective temperature)
        (L - luminosity)
estimated mass - m ⎫ several values, if the position of the
estimated age - t ⎬ star in the H-R diagram corresponds to
probability   - w ⎭ several evolutionary tracks

4. There exist also determinations of values m, t and w for stars of the Woolley catalog [4] brighter than $M_V \leqslant 7^m$ and having $\frac{\Delta\Pi}{\Pi} \leqslant 15\%$ (for about 400 stars).

5. All these data are kept in the computer's memory and can be delivered on punched cards or taped lists of data.

References

1. Paczynski B., Acta Astr., 20, 43 (1970).
2. Iben I., Ap. J., 141, 993 (1965); 142, 1447 (1965); 143, 483 (1966); 143, 505 (1967); 143, 516 (1967); 147, 624 (1967); 147, 650 (1967).
3. Varshavsky V., Tutukov A., Nauch. Inf. Moscow 23, 47 (1972); 26, 35 (1973).
4. Woolley R., Epps E.A., Penston M.J., Pocock S.B., R. Obs. Ann., No. 5 (1970).

6. Nautshnye informatsii of the Astronomical council of the USSR Ac. of Sci., Moscow (Russian with English abstracts) No. 21, p. 58, 68, 1972, No. 31, p. 113, 1974.

7. -

8. Punched cards, taped lists of data.

9. Our data on punched cards or taped lists are available through interchange.

10. We use the computer M-222. Two persons work on this particular subject. The group is linked to the Computing Center of the Astronomical Council in Zvenigorod.

EXISTING DATA CENTERS AND THEIR FUTURE ROLE

W. A. Baum

Planetary Research Center
Lowell Observatory
Flagstaff, Arizona 86001, U.S.A.

Two IAU data centers, one at the Lowell Observatory and the other at Meudon, were organized in the early 1960s for the purpose of establishing a collection of planetary data and images accessible to all qualified scientists. These centers trace their origin to an IAU resolution sponsored by Commission 16 in 1961. Ours at the Lowell Observatory, known as the Planetary Research Center, was also recommended by the U. S. Lunar and Planetary Missions Board and by the U. S. Space Sciences Board Panel on Earth-Based Planetary Astronomy. Since its inception, our Center has been supported by a grant from NASA Headquarters.

As far as ground-based planetary observations are concerned, the principal resource material not already available in published literature was initially in the form of planetary photographs. Our collection (Baum, 1973) now includes about 118,000 different image sequences, with a total of nearly two million planetary images. This ground-based collection is, of course, now being augmented with copies of spacecraft images.

On the basis of our experience, we believe that a data center will succeed best if it is operated by a staff that is actually engaged in its own research. Guest investigators often need more than a data library or a plate collection. We find that they need to confer with (and sometimes collaborate with) experienced colleagues. And if there is a resident research staff, there will also be instruments and computers for the assessment of data. At our Center the use of the collection by guest investigators alone would tend to utilize less than its full potential. In an effort to derive more science from our (and NASA's) investment, the emphasis of related work by our staff therefore shifted

C. Jaschek and G. A. Wilkins (eds.), Compilation, Critical Evaluation, and Distribution of Stellar Data. 281-283.
Copyright © 1977 by D. Reidel Publishing Company, Dordrecht-Holland. All Rights Reserved.

long ago from the enlarging of the collection to the extraction
of scientific results from it.

Astronomical data centers are likely to become increasingly
involved with images, whether they wish to or not, because the
conventional distinction between images and data is rapidly dis-
appearing.   In the past, an astronomical image was typically
represented by a distribution of photographic grains on a plate
or film.   Today, it may alternatively be represented by an array
of numbers that can be processed in a computer and stored on
magnetic tape.   Moreover, a digital image can be translated into
a photographic image, and vice versa, with relative ease.

A digital image is not necessarily a secondary product
created by microphotometrically scanning a photograph.   Increasingly, astronomical images are being recorded at the telescope
with scan-readout (television-like) detectors, including those
that count individual photoevents and thereby provide photometrically linear response (for example, Boksenberg and Burgess, 1972).
As dramatically demonstrated by recent Viking orbiter pictures of
Mars, a digital camera system (Carr *et al.*, 1976) can also provide
images of excellent quality.

It has been suggested that data centers might store digital
images in raw form so that users of the centers can apply their
own favorite computer processing programs, but there are good
reasons for planning differently.   This question is of immediate
importance for deciding on the storage of recent spacecraft images
and also of future importance for ground-based astronomical images
as digital methods become more common.   A digitally processed
image differs from a raw image in somewhat the same sense that
reduced data (as stored by a data center) differ from raw data
(original observations). Instrument peculiarities and aberrations
are corrected for, and calibrations are introduced.   This process
tends to be rather complex and to require intimate familiarity
with each particular imaging system.   The best plan will ordinar-
ily be for data centers to store digital images in at least a
partially processed form.

In summary: (1) An astronomical data center that is only a
data library may be less well utilized than one that is also an
equipped research center. (2) The digital recording and processing of images seems likely to have a major impact on the future
functions of data centers.

REFERENCES

Baum, W. A. The International Planetary Patrol Program: An
    Assessment of the First Three Years. *Planetary and Space
    Science* 21, 1511-1519, 1973.

Boksenberg, A., and Burgess, D. E.   An Image Photon Counting System for Optical Astronomy. *Advances in Electronics and Electron Physics* 33, 835-849, 1972.

Carr, M. H., Masursky, H., Baum, W. A., Blasius, K. R., Briggs, G. A., Cutts, J. A., Greeley, R., Guest, J. E., Smith, B. A., Soderblom, L. A., and Wellman, J. B.   Preliminary Results from the Viking Orbiter Imaging Experiment. *Science* 193, in press, 1976.

THIRD GENERAL CATALOGUE OF MK SPECTRAL CLASSIFICATIONS

William Buscombe

Northwestern University
Evanston, Illinois 60201

For the convenience of astronomers who need information concerning certain stars but prefer not to search through hundreds
of research articles in the hope of finding it, a file of MK
spectral types and UBV photoelectric photometry is maintained at
Dearborn Observatory. A generous donor made possible the publication in 1974 of a listing of approximately 17,000 IBM cards of
data compiled by Pamela M. Kennedy and William Buscombe during
1963-1973.

A sequel nearing completion will include for a similar number
of additional stars:

col. 1- 3 indication of multiplicity from spectroscopic
observations
6-11 designation in the order of preference:
Henry Draper Catalog,
Durchmusterung number,
Chart number for stars in clusters,
Variable star name
14-20 right ascension 2000
23-28 declination 2000
32-45 MK spectral type, luminosity class, peculiarity
48-50 reference number for classification
53-57 V magnitude
59-63 B-V colour index
65-69 U-B colour index
72-74 reference number for photometry
75-80 notes in the order of preference:
- indication of multiplicity from photometric
observations,

C. Jaschek and G. A. Wilkins (eds.), Compilation, Critical Evaluation, and Distribution of Stellar Data. 285-286.
Copyright © 1977 by D. Reidel Publishing Company, Dordrecht-Holland. All Rights Reserved.

                - variable star name, or alternative desig-
                  nation,
                - name of cluster for proven member stars

    The selection of objects for inclusion is based on recently
published journals, with special attention to nearby stars, binary
systems, variable stars, x-ray sources, supergiants, members of
galactic star clusters and the Magellanic Clouds.

    It is a pleasure to acknowledge the assistance of Michael
Allen and Mark Lewis in preparing this compilation.

———————

    NOTE: Since Dr. Buscombe was unable to attend the meeting,
only this abstract is provided.

THE FUTURE ROLE OF DATA CENTRES IN ASTRONOMY[1]

G. A. Wilkins

Royal Greenwich Observatory, Herstmonceux Castle, U.K.

There are two main modes of data-centre operation – the passive and the active. In the passive mode the data are received, catalogued and stored, and some are later copied and distributed in response to specific requests. In the past the passive mode has been represented by the many printed volumes of observational data in astronomical libraries, but the techniques of data acquisition are now such that it is often no longer practicable, even if it were desirable, to print the very large amounts of observational data that are now produced by modern instruments. In the active mode of data-centre operation the centre collects data that it considers to be useful, then evaluates, combines and analyses them, and finally publishes the results of this work. In the past the active mode has been common in astronomy, and the general catalogues of stellar data are examples of this mode of operation, but each one has required many years of effort. Modern computer systems can, however, store large amounts of data and can display, manipulate and copy them very quickly; it is now possible to combine data of many different kinds and to analyse them together, with the prospect of giving new knowledge about the systems being studied.

Astronomers have to decide how best to proceed in these new circumstances. Should we merely encourage the development of common standards for the publication of data on, say, magnetic tape, so that each institution can obtain the data that its astronomers wish to study? Or should we aim to develop a small number of very large, passive data centres which will make observational data available on request, and perhaps provide computer facilities to

---

[1] See also Report of discussions, section 7.

C. Jaschek and G. A. Wilkins (eds.), Compilation, Critical Evaluation, and Distribution of Stellar Data. 287-288.
Copyright © 1977 by D. Reidel Publishing Company, Dordrecht-Holland. All Rights Reserved.

visiting astronomers? Or should we aim to develop a network of
specialist, active data centres that will cooperate together, and
aim to build up a communal store of evaluated data and derived
results? Or can we find some combination of these three approaches
that will fit the characteristics of the data and of the astronomers who will use them?

ASTRONOMICAL DATA COORDINATION: A PERPETUAL TASK

W. D. Heintz, Vice-President, IAU Commission 5

Department of Astronomy, Swarthmore College,
Swarthmore, PA 19081, USA

The conference now about to be concluded has presented a
wide spectrum of coordinate efforts, of accomplishments, projects,
and viewpoints. As is only natural and even purposeful in a first
colloquium of this scope, there are also some loose ends which we
hope to tie together later. Many experienced scientists found it
worth participating in the subject which is more organization
rather than primary research, and this is a very encouraging feature. Most of us will have deplored, in one context or other, the
lack of reliable data compilations; yet not much could be done
before efficient computer and storage facilities were at hand,
and we may also have nourished the hope that someone else would
undertake the tedious and often thankless job of compilation. We
are grateful that the invitation to convene the first Numerical
Data conference came from the Strasbourg Observatory, where a large-
scale and multi-purpose program has been successfully developing
in recent years.

Bibliography has had a long history in astronomy and also in
the IAU, along with the other organizational work in the commissions on ephemerides and on telegrams. Under the increasing wave
of publications the tasks have greatly amplified and changed. How
to channel the flood of results, whether digits or prose? Several
new and viable working groups have formed: our consultant friends
among the astronomical librarians, the group of journal editors,
and here the specialists on numerical data. The groups represent
astronomy in multi-disciplinary organizations like the ICSU
Abstracting Board, the International Federation of Library Associations, and the CODATA committee, and they participate in work at
the high level of UNESCO and ICSU. I believe that science-funding
agencies in several countries are well aware of the increasing

C. Jaschek and G. A. Wilkins (eds.), Compilation, Critical Evaluation, and Distribution of Stellar Data. 289-291.
Copyright © 1977 by D. Reidel Publishing Company, Dordrecht-Holland. All Rights Reserved.

importance of secondary sources. Not only the primary data acquisition but also well-considered projects of compilation and
retrieval should be found worth having some funds channelled into
them.

More extended fields of science are still harder hit by the
information explosion, as the decreased printing of full-length
papers and the creation of synoptic journals show. Astronomy is
also only marginally concerned with the intricate problems related
to the commercial use of data. Yet we have problems of our own,
for instance the long lifetime of many astronomical results, since
many observations - unlike experiments - cannot be repeated at
will, and a sizeable part of observing effort is for use in some
future time. Results on a vast number of objects are to be kept
separate since their value is individual as well as statistical,
and moreover we have the peculiarly high fraction of publications
in non-periodical, non-commercial forms with limited accessibility.

The enhanced responsibility carried by the author of a secondary source has been repeatedly mentioned, that is, with regard to
the correctness of the data and - as the case may be - to their
completeness or competent selection. Researchers are both producers and users of stored data, and the expedient and unambiguous
presentation of data is needed for both ends of the supply line.
Other items falling into the realm of the data trade are: to agree
on standard formats for expedient processing, to find out what
constitutes a feasible amount of redundancy and duplication, to
advise on the extent or format in which data should continue to
be published in traditional form, and perhaps also on the use of
units and symbols. A special point may be made about the instruc-
tional value of the knowledge of sources, embodied in C. Jaschek's
clear introduction to abstracts, catalogues, and data storage
services this morning. Advanced students find it rewarding if at
least one seminar period is devoted to the subject: How to use a
library. An early ingrained know-how about records will save the
prospective young astronomer countless hours of frustrating
searches later on.

It was the work of the Strasbourg Observatory staff and their
numerous helpers which rendered this conference a successful and
pleasant event, in both the scientific and the social programs.
V. Abalakine has already given a comprehensive tri-lingual catalogue at the Closing Dinner; so let me just join with our deeply
felt thanks to our hosts. Let me also mention the indefatigable
activity of our chairman, G. Wilkins. Compare what he has accomplished from the tentative beginnings in Sydney 1973 up to today.

Used though we are to work with printed catalogues, future
data transfer surely will increasingly consist of communication
with data centers. Science shall not be suffocated in the quick-

sand of printed paper. Yet this meeting and particularly this morning's session will have made it abundantly clear that the preservation and utilization of what we already have is an indispensable foundation of high quality and efficiency of future research.

CONCLUDING REMARKS

R. H. Garstang

Joint Institute for Laboratory Astrophysics,
National Bureau of Standards and University of Colorado
and Department of Physics and Astrophysics, University
of Colorado, Boulder, Colorado   80309   U.S.A.

When listening to the papers which have been presented I was
struck by the feeling that much more effort is being devoted to
the problems of the acquisition and handling of vast quantities
of data than to the evaluation and presentation of the data, a
situation which has in the past applied to spectroscopic data as
well. The list of Data Centers which Dr. Jaschek presented
attests to the efforts now underway to collect data. I would
therefore urge strenuous attempts to critically evaluate collections of data whenever such evaluation is meaningful and feasible.
I believe that many of our colleagues expect us to tell them a
'best' value for any particular datum, with, if possible, an
indication of whether it is considered trustworthy (by a probable
error or a quality index). I believe that this principle of
giving the best representative values underlies much of the
success of C. W. Allen's book Astrophysical Quantities as well
as of compilations such as the National Bureau of Standards
spectroscopic publications. I would also emphasize the continuing value of bibliographies, at least in cases when the
original data are published in widely scattered articles and
until the publication of a comprehensive compilation. I believe
that good bibliographies on topics of special interest could go
far to meeting the casual needs of many users of our data.

Many speakers emphasized the importance of accurate and
clear descriptions of tapes and programs, and I would add all
published papers. The readership for many of our papers has
expanded to include physicists and geophysicists who are not,
and cannot be expected to be, familiar with our astronomical
traditions. The problem is acute in matters of notation and

C. Jaschek and G. A. Wilkins (eds.), Compilation, Critical Evaluation, and Distribution of Stellar Data. 293-294.
Copyright © 1977 by D. Reidel Publishing Company, Dordrecht-Holland. All Rights Reserved.

units. We had some discussion as to the desirability of using SI
units, which are unfamiliar to many astronomers but which appear
to be taught with increasing frequency to our students. I doubt
that we need to make a decision now as to whether we should all
adopt SI units. But I do think that we should take great pains
to state very clearly the units we use, and whenever we use units
unfamiliar to the wider scientific community we should state
their equivalents near the beginnings of our papers and as footnotes to tables of data. I think that that alone would help to
improve communication with colleagues in related fields. I also
think that care with the references at the ends of our papers is
worthwhile. One should not over-abbreviate the titles of publications. It is helpful to quote the last page as well as the
first page in a journal reference if the Editor will allow it --
this tells people who do not have the journal what it is they
have to order. Catalogues and privately published documents
should if possible have authors names on the title page, not
just the name of their institution: it has been my experience
that in large libraries the author index is far easier to use
than the subject indexes. Finally one should give as complete a
reference as possible for older publications: there are many
newer astronomy departments and research institutions which
possess hardly any of the older literature, and a complete
reference facilitates obtaining a copy from elsewhere.

Finally I would urge continuing efforts to make as much as
possible of our data available in forms accessible to non-
specialists. Clearly many large compilations cannot be printed
in book form. But I believe that there will still be a need,
for at least the foreseeable future, for publication in book
form in selected cases. An outstanding case is the Bright Star
Catalogue: this has been of great utility, not least because
of its ready accessibility. Indeed it is quite impossible for
Miss Hoffleit to know all the users of her splendid compilation.
My personal experience with spectroscopic data has been salutory:
when I looked up citations to my own papers in Science Citation
Index, I found quite a number of citations by authors unknown to
me, working on problems unfamiliar to me, and publishing their
papers in journals I would not normally read. I could not
possibly have known what use would be made of my data. I have
also spent much of my time over the years trying to help
colleagues and students with spectroscopic problems, and I have
discovered just how few industrial laboratories and small aca-
demic institutions have the astronomical journals, government
documents, and other sources of spectroscopic data. In short,
I would emphasize the importance of making our data available in
as accessible a form as possible, and so doing what we can to
ease the lot of our fellow scientists.

REPORT OF DISCUSSIONS[1]

G. A. Wilkins

Royal Greenwich Observatory, Herstmonceux Castle, U.K.

SUMMARY

This report summarises the discussions on the topics of the five invited papers, and on other matters of general interest that were raised by the contributed papers. The principal headings are: standards for datafiles; the influence of acquisition and processing techniques; the critical evaluation of data; the designation of astronomical objects; the distribution of data; survey of facilities; the future role of data centres; and recommendations on IAU activities.

1. STANDARDS AND METHODS FOR DATA HANDLING

1.1. Database management systems

In the first invited paper M.S. Davis (1) remarked that astronomers were amongst the first to use punched-card machines and computers for processing large data sets, such as the General Catalogue, but astronomers have not yet taken advantage of the facilities of data management systems that have been developed for other purposes. He made the point, which was later stressed again by Westerhout and others, that each datafile which is to be made available for processing by computer techniques should, like a good computer program, be self-documenting; that is the file should contain a full description of the data that it contains, and information about format, sources, precision, ranges of validity, peculiarities, formulae, etc., should be included. Such

---

[1] Papers are referred to by authors and numbers as shown in the table of contents.

C. Jaschek and G. A. Wilkins (eds.), Compilation, Critical Evaluation, and Distribution of Stellar Data. 295-316.

information is now often, but not always, given in the prefaces
and footnotes of printed catalogues, and is usually omitted when
the catalogue is transcribed for computer processing; the new
techniques of data management allow for the inclusion and retrieval
of such information.

Davis, in reply to a question by Walter, said that the costs
of file maintenance (deletions, additions and corrections) depend
on the file structure; they can be enormous for a sequential file
but modest for a list structure, and the need for file maintenance
must be considered in the planning phase. When catalogues are
transcribed from printed to computer-readable form, it is more
efficient to omit computable quantities from the transcription;
it would in any case be necessary to recompute them for checking
purposes.

In a later paper [Harten and Spoelstra (11)], Spoelstra
described the data formats used in connection with the Westerbork
radio telescope, with special reference to the special format for
"transport tapes", i.e. for magnetic tapes which are to be sent to
other institutions. He considered that there should be standard
formats and a restricted set of tape characteristics for such tapes;
in particular, he suggested that the structure of the descriptive
label at the beginning of a datafile should be the same for all
branches of astronomy. He also suggested that the label should
include the name of a person who could be contacted about the
file, as well as the technical parameters of the instrument and
the descriptions of the contents and format of the data.

1.2. Errors in datafiles

M.S. Davis (1) also drew attention to the nature of errors,
to the techniques for their detection and to the vital importance
of ensuring that datafiles are not corrupted or lost during use
and maintenance (or by unauthorised alteration). R.J. Davis commented that errors occur during the initial collection, reduction
and publication of data, and can only be detected later when they
conflict with later, independently acquired data; bibliographic
and personal problems can arise if such errors are corrected uni-
laterally. M.S. Davis considered that the author and the data
centre that distributed the file should be informed, and other
users should be notified through the publication of errata; problems need only arise if an author is negligent or obstinate. In
a later paper Bidelman (17) pointed out that errors are sometimes
not treated properly, and many authors prefer to allow their errors
to go unnoticed. Dixon stressed that any errors found in a catalogue should always be reported to the author; it is unfair to him
if this is not done.

In this connection Jaschek requested that details of errors
in datafiles be sent to the Stellar Data Centre at Strasbourg and
he offered to publish them in the Information Bulletin of the
Centre. Duncombe suggested that the U.S. Naval Observatory would
publish errata lists in its Circulars on the data available in
machine-readable form. R.J. Davis suggested that each distribution
centre should send details of errors to those to whom it had sent
copies of the datafiles.

Collins asked whether experience had shown that optical-
character-recognition machines had cut down the amount of proofreading required when transcribing printed material for computer
processing. M.S. Davis replied affirmatively, but pointed out
that it will be a long time before it will be possible to read all
documents automatically; in the meantime he recommended the use,
where possible, of key-to-tape or key-to-disc systems, rather than
the intermediate use of punched cards.

1.3. Other aspects

Many of the later papers and discussions were related to the
need for standards for improving the quality of the data and for
standards for facilitating the distribution of data. Section 3.2
on the presentation of data in the primary literature and section
5.1 on the publication of data on magnetic tape are of particular
relevance.

Wilkins drew attention to the Task Group on the methodology
for handling space- and time-dependent data that had been set up
by CODATA on the recommendation of its Advisory Panel on the
Geosciences, of which he is Chairman. The Chairman of the Task
Group will be a geographer, Dr. R.F. Tomlinson, and it will contain
scientists of many disciplines, including astronomy, who will be
experts in the handling of data; it will seek to identify common
problems and the possibilities for wider use of solutions developed
in one discipline. Wilkins expressed the hope that astronomers
would both contribute to, and benefit from, the work of this Task
Group.

2. INFLUENCE OF ACQUISITION AND PROCESSING TECHNIQUES

2.1. The next 10 years

The principal theme of Westerhout's invited paper (9) was the
necessity for astronomers to prepare for the great increase in the
volume of data that will become available in, say, the next ten
years because of the increasing sophistication of the instrumenta-

tion that is now coming into use for ground-based and space astronomy. On-line minicomputers will be used for the initial processing
of the data, but even so the data for one night's work could fill
one reel of magnetic tape. On the other hand, new computer devices
will allow greater packing densities for the information and the
new interactive techniques for data processing will lead to new
uses for the data. He suggested that the IAU should set up a small
panel to monitor the developments in computer hardware and techniques and to make recommendations about standard media and formats
for data distribution. This suggestion was discussed again in the
final session (section 8).

In the ensuing discussion, Nandy commented that the photographic plate is an economic storage device, and a plate can be
measured quickly using new automatic measuring machines, which can
reach the 24th magnitude. Westerhout agreed, but pointed out that
photographic plates are not linear and that there are features in
the universe which cannot be resolved by photographic plates; an
array detector will be able to distinguish objects which have very
large differences in intensity even when they are very close together. Baum added that there can be no doubt that array detectors will soon be providing much of our astronomical data and will
therefore have a major impact on the operation of data centres;
but not all array detectors are photon-counting devices with direct
digital output. Some arrays, including change-coupled devices,
require analogue-to-digital conversion of the signal. Photon
counting arrays, which directly provide a digital output with
precise linearity, are severely limited in counting rate and are
therefore suitable only for extremely faint sources; although
improvements in photon counting arrays are foreseen, a major
breakthrough will be required to fully overcome this limitation.

In his talk during the final session Baum (34) returned to
the point that the conventional distinction between images and
data is rapidly breaking down. Images now often exist in digital
form, and this may even be the primary form (e.g. for planetary
spacecraft and the Space Telescope). These digital images may be
processed by computer to remove faults and apply calibrations;
this initial processing is too costly and too specialised to
repeat. This point of view was supported by Underhill, who remarked that a long learning process will be required, and by
Westerhout who considered that it would be unrealistic to store
raw, unprocessed data, but that it will be necessary to decide
whether to hold both reduced data and the corrections for external
factors that have been applied to the processed data to reduce
them to standard form.

Mistrik questioned whether it would be more economical to
represent the data by some appropriate mathematical expression
(for example, by the coefficients of a polynomial), but Westerhout

considered that this would be limited by the presence of noise in
the data. In his talk on data processing for space-based astronomy, Mistrik discussed in more detail the problems of where the
data should be processed and gave an example of the rapid accumu-
lation of data in a space project.

2.2. Other facilities

In presenting a joint paper with T.A. Nagy on retrieval techniques and graphics displays, Mead (24) gave an example of how
overlays for use with Palomar sky-survey plates can be produced
from a database formed by combining four catalogues and sorting
the data according to plate area. The programs for this, and other
tasks, have been written for an IBM 360/65 computer. Baum and
Underhill also mentioned facilities for image processing and for
interactive graphics that are available at the Jet Propulsion
Laboratory and the Goddard Space Flight Center.

In describing the facilities offered by the Stellar Data
Centre at Strasbourg, Ochsenbein [Ochsenbein, Egret and Bischoff
(5)] drew attention to the way in which the observatories in France
are linked to the IBM 360/65 computer at Meudon and hence have
direct access to the data files of the Centre. Apart from this,
there was, however, very little discussion of current applications
in astronomy of interactive techniques.

Hauck later drew attention to the facility by which data may
be printed on microfiche rather than on line-printer listings.
Jaschek saw this as a way of making datafiles available to astronomers in institutions which did not have computing facilities, as
well as a way of avoiding the high costs associated with printing
by conventional means; one microfiche can contain up to 220 pages
of text and costs less than 2 US dollars.

3. THE CRITICAL EVALUATION AND PRESENTATION OF DATA

3.1. Construction of catalogues

In an invited paper on the critical evaluation of data
Underhill [Underhill, Mead and Nagy (15)] discussed the particular
problems that arise in connection with catalogues that are available on magnetic tape. Many of these catalogues do not provide
the information that is required if the user is to be able to judge
the quality of the data. Even where the catalogue is derived from
an earlier publication it is necessary that the information about
methods of observation and reduction be included with the data,
since users may not have access to the original publications,

especially if they work in a new institution that does not have a
comprehensive library. She stressed that the error estimates for
the data, and their basis, should be clearly given, and she gave
specific examples of other relevant items that are necessary to a
proper evaluation of the quality of astronomical data.

In the discussion Jaschek mentioned the situation in photometric photometry, where practically all the measurements before
1950 are lost because the authors did not specify precisely enough
the transmission band used; the Lausanne group has, however, published the bands of most of the modern photometric systems. Gliese
commented that details of instrumental techniques are necessary in
the compilation of astrometric data to form a fundamental catalogue. The Astronomisches Rechen-Institut has asked that the
observational catalogues to be used in the preparation of FK5
should be supplied on punched cards or magnetic tape, but it is
not sufficient to be told that the catalogue was observed, say,
"semi-absolutely"; rather the methods of determining azimuth,
flexure and the magnitude equation should be stated explicitly.

In another session Dixon commented that it is still desirable
to include computable quantities in the printed catalogues for the
benefit of those who do not have computers at hand; those having
access to large computers should not assume that everyone else is
also well-equipped; the published material should be aimed at the
widest audience.

The problems encountered in the formation of a homogeneous
catalogue of photometric data were discussed by Nicolet [Nicolet
and Hauck (16)]; in reply to a question by Nandy he stated that
the observations were weighted according to their dispersions.

The following abstract, which was submitted by Mrs. K.
Haramundanis (Smithsonian Astrophysical Observatory), was explicitly considered during the discussion on the role of data centres,
but it is also relevant in this context.

The most useful, reliable, trustworthy and effective
compilations of astronomical data are prepared by specialists in their field, since critical evaluations are
significantly more useful than lists. To assist the
integration of critical evaluations, an international
standard of identifications must be established and adhered to. Data should be disseminated in both computer-
accessible and book form. The work and expense of preparing compilations of critically evaluated data could
be effectively shared if each data centre were to be
responsible for a particular sky area only.

3.2. Presentation of data in the primary literature

During the discussions on the papers by Davis and Underhill
it became clear that there was general agreement that many papers
containing original data do not contain sufficient information to
permit readers to assess the quality of the data and results.
Wilkins drew attention to the CODATA "Guide on the presentation
of experimental data in the primary literature" (CODATA Bulletin
No. 9, 1974) and suggested that it would be worthwhile preparing
a corresponding guide for astronomical data obtained by observations. Lortet pointed out that it would be necessary to persuade
editors to accept that such details should be given. Wilkins
agreed and commented that the guide would be addressed to editors
and referees as well as to authors. He expressed his disappoint-
ment that astronomers were not changing over to the International
System (SI) of units more quickly, especially as the delay will
cause unnecessary difficulties for those who are now being taught
physics, chemistry, etc., in terms of SI units.

Van Altena drew attention to the multiplicity of units for
angular measure by asking what is the SI unit for angle. Wilkins
stated that it is the radian but the other units are recognised.
McCarthy suggested that the change to SI units would be facilitated by the insertion of conversion tables in the next edition
of Allen's Astrophysical Quantities; Wilkins said that he would
like to convince Professor Allen that he should use SI units
throughout the next edition. He also referred to the usage in
the CODATA guide of the terms "imprecision" and "inaccuracy" for
the measures of the internal and external errors of measurements
and results; he considered that their logical appropriateness
outweighed the disadvantage of their present unfamiliarity.

Westerhout agreed with the desirability of giving estimates
of both "imprecision" and "inaccuracy" but went on to ask how
many astronomers are represented on the CODATA panels that propose the new terminology; the long-established practices of astronomy should be taken into account by such international panels.
Wilkins replied that, as far as he was aware, no astronomers were
members of the panel that drafted the CODATA guide on experimental
data, but that CODATA was encouraging the preparation of other
guides that were suited to their fields of application. Underhill
considered that astronomers were too few in number to carry such
weight, and that it will continue to be necessary to be aware of
internal errors, external errors, random errors, and systematic
errors, whatever their "official" names may be.

Luyten expressed the opinion that the abbreviation pc for
parsec is absurd, and that analogy with other abbreviations suggests that the abbreviation should be ps. (This would, however,
mean picosecond in the nomenclature of SI.)

4. THE DESIGNATION OF ASTRONOMICAL OBJECTS

4.1. Current problems

Several of the papers presented at the Colloquium were concerned in whole or in part with the problems caused by the multiplicity and ambiguity of the designation systems for astronomical
objects lying outside the solar system. There were also many
other comments on these problems and several suggestions for their
amelioration. As the following notes and selection of comments
will make clear, there was no general agreement as to the designation systems that ought to be adopted.

Mermilliod's paper (2) on the principles of a coded numbering
system for photometric data and its application to open clusters
opened up the subject. In response to comments and questions he
stated that the system did not introduce any new designations; it
provides a translation of existing identifications for practical
use in machine-readable files; in principle it allows for the inclusion of faint stars which are identifiable only by provisional
numbers on charts; where an open cluster has two or more different
numbering systems, one of them is adopted as a base and transformation tables are constructed for the others.

Ochsenbein [Ochsenbein, Egret and Bischoff (5)] described the
catalogue of stellar identification that is available at the
Stellar Data Centre at Strasbourg. This originally provided cross-
references between the designations used in half-a-dozen or so
catalogues but it had since been extended and now covered some
400 000 stars. The formation of this cross-index had revealed
some errors in the original catalogues. Underhill [Underhill,
Mead and Nagy (15)] and Mead [Mead and Nagy (24)] both referred
to a comparable cross-index that had been started independently
at the Goddard Space Flight Center; the two institutions are now
cooperating on this work.

Collins (4) drew attention to the ambiguities in designations
that he had found while forming a bibliography of astronomical
catalogues published since 1951. He suggested that there should
be control of new designations. He had given W175 as an example
of a designation for several different objects, and Luyten added
that it also referred to the proper motion star Wolf 175. In
reply to a question, Collins stated that the computer file is not
yet closed, but he hoped that the bibliography would be published
by INSPEC before the end of 1976.

Spite (7) discussed the same problem and suggested that each
author should be expected to give the full reference to each original catalogue that is not included in a list of designations that

would be recognised by the IAU. Jaschek commented that it is only
necessary to convince the editors of the leading astronomical
journals of the desirability of this proposal. If they refused
to publish papers which did not provide correct identifications
the problem would soon be solved. Mead considered that the editors
should require authors to provide coordinates for all objects
described in a paper.

Underhill made the suggestion that, as many of the early,
small catalogues are in obscure publications that are not available in newer libraries, it is desirable that such catalogues be
transcribed into machine-readable form. Spite thought that this
could be done with a relatively small amount of extra work and
that it would be better to give a reference to where such a catalogue is now available, or to a newer and better catalogue, rather
than to an old publication that is now out of print.

There were other suggestions for short-term remedies. For
example, Luyten urged that if a star is contained in the Bonn
Durchmusterung (B.D.) or the Cordoba Durchmusterung (Co.D.) no
other designation should be used; these catalogues contain about
one million stars and are now almost universally used. He later
expressed his concern about the coining of new designations for
stars that have been in the literature for a long time; for example, new designations are often given when a star is observed by,
say, a spectroscopist even when it has previously been designated
as a white dwarf; some such stars now have five or six different
names. Bidelman suggested that there should be a recognised list
giving the order of preference for the use of designations when a
star was listed in more than one catalogue; in particular he considered that common names should be used in preference to catalogue
numbers. Others felt that it would be difficult to get general
agreement to such a list since the usefulness of a catalogue
depends on whether it contains the information required for the
application.

4.2. Use of coordinates in identifiers

There seemed to be general agreement to the view that for the
identification of data held in computer files a system based on
the use of the coordinates of the object would be appropriate, but
there were differing views as to the desirability and practicabil-
ity of having a single system for all astronomical objects outside
the solar system and as to the particular form that the identifier
should take. The following notes give a selection of these views.

In his paper, Eichhorn (3) suggested that the identifiers
should be formed from galactic rather than equatorial coordinates.

M.S. Davis questioned whether it might be better to use an inertial
frame defined with respect of extra-galactic objects rather than
a galactic frame which is defined with respect to the equatorial
system. Westerhout commented that any fixed frame could be used
and that the necessary coordinate conversions can be carried out
at tape-reading speeds. Dixon pointed out that galactic coordinates are not particularly significant for extra-galactic objects.
Van Herk suggested that space astrometry will provide an alternative system. Bidelman considered that it is more important that
everyone should use the same equinox. Heintz drew attention to
the imminent changes in the equinox and precession coefficients
of the fundamental system. In his replies, Eichhorn emphasised
that whatever the choice of system it will always be necessary
to make transformations between the moving equatorial system in
which we observe and the chosen system, and this transformation
will be uncertain for coordinates of high precision.

Dixon later returned to the question of the choice of the
standard epoch for use in catalogues. He argues that the proposal
to be considered at Grenoble for the adoption of 2000.0 as the new
standard epoch would involve much unnecessary work in changing
existing catalogues and that therefore the epoch of 1950.0 should
be retained indefinitely. Wilkins pointed out that the proposal
was specifically related to the standard epoch of the new fundamental catalogue FK5 and of new planetary theories. Bidelman commented that the adoption of 2000.0 for the fundamental catalogue
did not preclude the continued use of 1950.0 for the epoch of
coordinates intended primarily for identification purposes; it
will be necessary to continue to provide the appropriate precession
coefficients for transformation between 1950.0 and current epochs.
Westerhout supported this view and, on this basis, Dixon did not
press his point further.

In connection with the problem of identifiers for non-stellar
objects whose positions cannot be specified precisely, Wilkins
drew attention to a technique used in a geographical data project
in which the identifiers were formed by combining the degrees of
latitude and longitude, then the minutes, and finally, the seconds;
with this technique, the identifier need be no longer than the
known precision demanded. Underhill commented that we need a technique that can be used efficiently in the computer, and the view
was expressed by Mead that we should not perpetuate the use of
sexagesimal measure in datafiles for computer use. It may be
noticed that Eichhorn suggested publishing direction cosines, which
also have the advantage that they avoid the singularity at the pole
of a spherical coordinate system. McCarthy and others suggested
the magnitude should be included in each identifier since this
would help to guard against misidentifications.

There were divided views about the identification of stars
in photometric sequences, clusters, and other dense fields. At
present stars are often identified only by a number on a particular
chart. Bidelman suggested that it ought to be practicable to
determine coordinates for such stars with sufficient precision for
identification purposes, using, for example, Palomar sky-survey
charts; a central service could do this. He considered that the
coordinates should be determined so that the data can be integrated
into a general astronomical datafile. Hoffleit considered that
charts will continue to be useful and that such coordinates would
not be sufficient. Westerhout agreed on the need for charts and
considered that coordinates should be given to the highest possible
accuracy; they are important because they can be put in machine-
readable form and avoid the necessity for publishing finding
charts, which can be produced by computer when required. Argue
pointed out that some finding charts do not resemble very closely
what is seen at the telescope eyepiece; no difficulty over identification arises if the telescope has a setting accuracy of about
1"; when a telescope does not set to this accuracy it may be possible to set on an easily identifiable bright object, and to offset from this. He claimed that positions can be easily read from
survey plates to an accuracy of $\pm$ 3" using simply an eyepiece and
to 0".5 using a simple X,Y measuring machine. Baum commented that
the coordinates of stars belonging to a local magnitude sequence
will not provide the complete information needed for a user to
construct his own finding chart, because only a fraction of the
stars in a field are selected for photometric measurements; if the
author provides no finding chart, he will have to provide coordinates for many more stars than he measures photometrically.

5. THE DISTRIBUTION OF DATA

5.1. The publication of data

In his invited paper Hauck (23) considered the basic problem
of how astronomical data should be published in view of the increasing rate of acquisition. He considered that there will be
a continuing need for printed catalogues but that the basic support for them should be on magnetic tape. He described a scheme
whereby tapes may be refereed through the journal system – the
referee examines the description of the datafile that is published
in the journal while a cooperating data centre verifies that a
copy of the tape satisfies its specification and makes further
copies available on request. The scheme allows for an author to
get credit for his work, provides a reference and an adequate
source of information about the data, and should filter out datafiles that are based on poor observing and reduction techniques.
Such a scheme is offered by <u>Astronomy and Astrophysics Supplements</u>

in cooperation with the Stellar Data Centre at Strasbourg. Other
centres will be required if the scheme is successful.

Batten expressed his concern about the role of the referee;
the compiler of a catalogue may have been in touch with potential
users and be in the best position to judge the best format in
which to present the data; it would be unfortunate if a referee
with strong opinions could force a compiler to adopt a presentation that was later found to be less suitable. Underhill had
similar opinions and suggested that, since preparing and publishing a catalogue is very expensive, the refereeing should be done
on the basis of trial data. Hauck agreed that it is necessary to
define precisely the tasks of a referee (see note at end of section
5.4), but considered that his task is a very useful one and that
he should examine the realisation as well as the intention. Argue
favoured the scheme because of its implications for students; at
present the submission of data is not regarded as having the same
weight as the publication of a research paper, but the status of
data compilation would be enhanced if it were known that the data
had been submitted to a proper system of refereeing.

Collins reported a survey which had shown that the half-life
for the citation of stellar data is the same (5 years) as that of
the research papers themselves; he concluded that there is an
urgent need to make the magnetic tapes available in a very short
period. It is possible that the unexpected result of the survey
is due to the incorporation of the older data in later catalogues
or it indicates that astronomers do not, in general, search the
older literature for data.

5.2 Sources of information

Several of the papers were concerned with services and publications intended to provide information about where data is to
be found. Collins (4) described work on a bibliography of recent
catalogues, including quite short lists of data. Spite [Ochsenbein
and Spite (27)] described a computer-accessible bibliographic file
on stellar data that has been formed through collaboration between
the Paris and Strasbourg observatories. Walter commented that the
consistent usage of the abbreviations for astronomical journals as
adopted in Astronomy and Astrophysics Abstracts would simplify the
interpretation of the stellar bibliographic file, and he enquired
why a different notation is employed. Spite replied that the aim
is to save punching time and computer storage, and so to save
money; in his opinion such a limited number of abbreviations cannot produce serious ambiguities, and it is easy to build and publish a short table of the correspondences between the adopted and
official abbreviations. In reply to questions by Mistrik and
Straizys he stated that the contents of a file can be output to

microfiche using a standard package, and that the file includes
all variable stars that have a designation or coordinate; there
is no limitation on magnitude.

When presenting his survey of facilities Jaschek (33) drew
attention to the existence of many examples of abservations that
had been unnecessarily duplicated; Bidelman and others confirmed
this view that the data that are available are not used to the
full extent that is possible. There are many indexes to data that
are at present only available on hand-written cards; for example,
Bidelman has an index on stellar spectra and Wood (21) described
such a catalogue for eclipsing variables. There was general
agreement that these indexes, in particular, and there may be
others, should be transcribed to magnetic tape so that the information in them can be made more widely available. Parsons [Parsons
and Wray (14)] reported that work had started on the punching of
Bidelman's datafile.

Dixon (25) described the progress on the preparation of a
master list of non-stellar objects. The list will give both the
basic data and a bibliographic reference to the original source
for each object. No attempt has been made to evaluate the data,
and a single object may be given several times if the published
positions are different. Jaschek suggested that it would be useful to collect later bibliographic references for all objects, but
Dixon felt that there should be some consensus that this would be
necessary. Bidelman said that he felt that this would be useful,
and he had started to compile such a file; he asked how Dixon
handled NGC or IC objects that are actually parts of other galaxies. Dixon replied that the NGC, IC and all galaxy catalogues are
included in the master list; by looking at the positions and
angular diameters of each object it is easily determined if one
object lies within another; in many cases such objects are specifically identified in the remarks column. In reply to a question by Westerhout, he also indicated that there are plans to try
to combine all entries that refer to the same object, but this is
a major project in which many people can take part; the master
list is the first step.

Lortet (26) described a project for documenting and cross-
identifying non-stellar galactic objects, such as dark clouds and
reflection nebulae. She drew attention to the need for a set of
keywords for classification of such documents. Wilkins suggested
that the revised version of the Universal Decimal Classification
for astronomy (just prepared by a Working Group of IAU Commission
5) would provide a satisfactory alternative to the use of keywords;
it includes numerical codes to indicate, for example, the type of
object and the wavelength range of the observations.

Wood drew attention to the desirability of determining what
the Commissions of the IAU are already doing in the matter of data
dissemination, and gave the example of Commission 41 which distributes, twice a year, lists of references and research on individual objects. Some information on these activities has been
recently published in the Information Bulletin of the Strasbourg
Centre.

5.3. Requirements for new documentation services

In the discussion on Jaschek's survey of data centres (see
section 6.1) several participants drew attention to the need for
additional documentation services for special types of object.
Bidelman noted that a new fairly extensive list of shell stars,
together with a bibliography, will be included in the proceedings
of IAU Symposium No. 70 on Be and shell stars: he continued by
remarking that there are many other types of stellar object that
merit continued cataloguing - pulsars, X-ray stars, infra-red
stars, etc., to name but a few examples; further, galaxies, quasars
and extra-galactic radio sources deserve extensive documentation;
this does not appear to be being done at the present time, except
for the brightest objects of these classes.

McCarthy commented on the catalogues for emission-line stars;
the researches sponsored by the IAU at Prague had proved fruitful
and, with the publication of the Henize Catalogue of Southern Be
stars, this preliminary phase of the data collection is finished;
a single working catalogue integrating the best features of all
existing catalogues plus the new discovery lists can now be assembled at a data centre.

McCarthy also enquired about the present situation regarding
lists and catalogues of stellar associations and open star clusters. Jaschek replied that the activity has effectively ceased.
There is a great need for continuing the work on the Prague catalogue, but no persons seem to be willing to undertake this. Lynga
agreed that the Prague catalogue of open clusters had served a
useful purpose and that a similar bibliographic catalogue should
be kept up; he suggested that making a tape of the references
given in the IAU Reports on Astronomy by the President of Commission 37 would make a good start.

Westerhout, in reply to a request by Jaschek for information
about documentation work in radio astronomy, stated that he knew
of only Dixon's list of radio sources, Terzian's list of pulsars,
and a private abstracting service for the members of the CSIRO
Division of Radiophysics in Sydney; staff members scan all journals
weekly for articles on radio astronomy and related subjects, write
small abstracts with critical comments, and put them on tape.

Dixon stated that he receives a copy of the abstracts by airmail every two weeks or so, and that he believed the service was open to others. Jaschek commented that this is a nice example of a service which exists, but which is known only to the specialists.

Wilkins wondered whether the three principal abstracting services could assist in the collection of material for such specialised files. In some sciences the presence of data in an article is indicated by the attachment of a 'flag' or a 'tag' to the abstract. The extra effort required to identify data could, perhaps, be found by increasing the cooperation between the three services.

R.J. Davis suggested that the abstracting services, and institutional libraries, should receive information about catalogues that are available on magnetic tape in order that astronomers may be directed to these data sources by means of the common techniques for searching the literature. This would occur automatically if details of all such catalogues were given in the primary journals as suggested by Hauck.

5.4. The "ownership" of data

During the discussion on Underhill's invited paper, Jaschek drew attention to the personal problems that arise as a result of the availability of new techniques for copying, updating, and merging catalogues. The authorship of a printed catalogue is normally clear and well recognized, even though the authors' names may not be given in citations. There are, however, examples where datafiles have been copied and then updated, or expanded, and no recognition of the original author has been given when the new datafiles have been used. Underhill commented that this is one more reason why every catalogue, whether printed or only available on tape, should be accompanied by documentation that gives the sources of the data; proper credit must be given to those who prepare catalogues; machine-readable catalogues are a very necessary form of publishing; all rights envisaged in copyright law that are commensurate with the requirement of open transfer of scientific information must be preserved.

It may be remarked here that one of the tasks of the referee of a machine-readable catalogue (see section 5.1) should be to verify that the author has given proper recognition of the sources of the data. Further, Dixon suggested that when a catalogue is duplicated by a centre for use at other institutions, the original author should be sent a list of those to whom his catalogue has been sent; this will enable him to know how widely his work is being used and to whom he should send announcements of up-dates and corrections.

6. SURVEY OF FACILITIES

6.1 General survey of data centres

In the fifth invited paper Jaschek (33) gave a summary of the
replies that he had received in response to a questionnaire that
he had distributed earlier in the year. The survey revealed an
extremely wide variety of functions and policies. It showed that
work on non-stellar objects is less well organised than that on
stars, and this in turn is less well organised than that on solar
system objects and phenomena. It also showed that astronomers are
very slow to adopt new ideas and they do not utilise fully the
services available from the data centres. It is the intention
that the results of the survey will be published by CODATA, as
part of the production of a directory of data sources for science
and technology. The discussions on documentation services that
followed this paper are reported in section 5.3.

Jaschek had previously drawn attention to the setting up at
Potsdam of a centre containing a copy of the datafiles available
at Strasbourg; this is intended to facilitate the use of these
data by the countries of Eastern Europe; and other similar centres
could be established elsewhere. Duncombe suggested that the data
sets at USNO and RGO could also be made available to such centres;
he also drew attention to the increasing efforts to coordinate the
work of the centres, and drew attention to the service provided by
the Bureau des Longitudes, in Paris, which distributes cards giving
information about the availability of astronomical ephemerides and
catalogues.

6.2 Individual services and projects

During the Colloquium, many of the participants described
briefly data services and projects with which they are concerned.
Some matters of general interest that arose during these presentations and the ensuing discussions are mentioned in other sections
of this report. The following notes are therefore intended only
to record a few points of specialist interest; further details are
given in the papers themselves.

Terashita (30) described the activities of astronomical data-
systems groups in Japan; its first task was to find out what
datafiles were already available on tape in Japan; he felt there
is a need for a set of standards for coding data, such as spectral
types.

Abalakine read a paper by Polojentsev (31) on the computing
facilities at the Pulkovo Observatory; the suggestion was made

that astronomers should be provided with algorithms to compute
ephemeris data. Wilkins commented that the Nautical Almanac
Offices of the US Naval Observatory and the Royal Greenwich
Observatory were planning to publish such algorithms but he considered that there would still be a requirement for the publication of the ephemerides.

Worley presented the paper by Fiala and Seidelmann (32) on
the data services now provided by the US Naval Observatory; a new
Federal service for supplying copies of tapes might be recommended.

Gliese (6) described plans for the third edition of his catalogue of nearby stars; he expected it to contain about 1700 stars
with parallaxes greater than 0".045. Walter asked whether it
would be possible to label the stars that are emitting at radio
wavelengths; Gliese replied that this is under consideration.
Luyten stated that he had just completed a new catalogue of stars
with annual proper motions greater than 0".5; it contains 3600
entries, of which about 2500 are in neither the B.D. nor the
Co.D.; he hoped that it would be possible to prepare and publish
identification charts for all of these stars.

Walter (12) gave the preliminary results of a search for
observational data from radio interferometry that would be suitable for astrometric purposes.

Nandy (13) compared systems of stellar classification from
low-dispersion ultraviolet spectra as a preliminary to the preparation of a catalogue based on observations made in a satellite
experiment.

Parsons [Parsons and Wray (14)] discussed the preparation of
a catalogue of ultraviolet spectroscopic and flux data for early-
type stars; it will be based on merging new satellite data with
existing catalogues, and will be accompanied by a bibliographic
file.

Abalakine brought greetings from B.V. Kukarkin, who was
unfortunately not well, and then read the paper [Kukarkin,
Kholopov and Kireeva (18)] describing the work in the USSR on
variable stars, including the tentative plans for the fourth
edition of the General Catalogue on Variable Stars. Bidelman
congratulated the Soviet astronomers for their extremely fine
work on the documentation of variable stars and looked forward to
the new catalogue; he considered that the new tables of alternative designations would be very useful. In reply to questions
Abalakine said that he expected that the catalogue would be available on magnetic tape and that the introduction would be in both
Russian and English.

Argue (19) described the way in which he and Miller had prepared a catalogue of photometric sequences; the aim was to put
the "consumer" in contact with the data and the observer, and no
attempt was made to provide critically evaluated data; the observers could probably provide better current data than could be
found, less quickly, in the literature. The ensuing discussion
on finding charts is reported at the end of section 4.2.

Batten (20) then discussed the preparation of a catalogue of
spectroscopic binaries; in contrast, the aim here is to select
the best set of elements to represent the behaviour of each system.
The notes on the individual systems represent an important part
of the catalogue and justified the production in book form. In
reply to Mead, he said that the new catalogue will be available
on magnetic tape, but he did not consider that it would be practicable to include the notes. Westerhout argued that the notes
should go with the data on the tape, and could easily be included
in a separate alphanumeric subfile. Jaschek suggests that the
notes should mention any photoelectric observations made when
looking for eclipses; he stated that the group at Toulouse will
publish a bibliography of all orbits for each star.

Wood (21) described the history and use of the manual card
catalogue on eclipsing variables now maintained at the University
of Florida. He hopes to produce a new edition of the finding
list for observers and to transcribe the file to punched-cards -
this led to the suggestion that it is better to avoid the intermediate use of punched cards, and to edit the file via a computer
terminal.

Hoffleit [Hoffleit and Jaschek (22)] asked for suggestions
as to what changes should be made in the next edition; she said
that the catalogue aims to cater for users who do not have access
to computers. Slettebak asked that rotational velocities be considered. McCarthy expressed the hope that the temptation to
reduce the size of the type would be resisted since the catalogue
is used during the making of observations; Hoffleit replied: "Older
astronomers of my age group are well aware of this problem!" R.J.
Davis suggested that photometric data should be given in the ubvy
β system. Walter asked that both Córdoba and Cape Durchmustering
numbers should be given when both have been allocated.

Worley (28) described the arrangements for the maintenance
and use of the master copy of the catalogues for visual double
stars at the US Naval Observatory; there is an Index Catalogue and
an Observation Catalogue; old as well as new observations are
being punched to make the catalogue as complete as possible.
Gliese asked whether the catalogue included star pairs of common
proper motion even if their angular separations are large. Worley
stated that there is now no arbitrary separation limit, and previously omitted wide pairs are being recovered where possible.

Van Altena [van Altena, Hoffleit and Smith (29)] discussed the systematic differences between trigonometric parallaxes obtained at different observatories and described plans for the construction at Yale of a new general catalogue of trigonometric parallaxes. Gliese drew attention to Strand's view that parallaxes based on the use of astrometric reflectors are more likely to be free from systematic error and that the true parallax system would be about 0".002 greater than the Allegheny system. Van Altena commented that it will be very difficult to determine the zero point of the system of parallaxes until the causes of the systematic difference are identified. Modern parallax observations may define a reasonably good, uniform system of relative parallaxes, since there appear to be little, if any, differences between the parallaxes of, for example, the USNO and Yerkes.

7. THE FUTURE ROLE OF DATA CENTRES

Throughout the colloquium there were brief discussions on the future role of data centres and so the discussion during the final session took on the character of a short extension and review of the earlier discussions.

Baum (34) used the example of the planetary-photograph centres as the basis for his talk, which included the following points: data centres operate best if staff at the centre are engaged on research that uses the data; the conventional distinction between images and data is rapidly breaking down; we have to distinguish between raw data, processed data, and reduced data; and the initial processing is too costly and too specialised to be repeated (see also section 2.1). In addition we have standard data that are obtained by combining together data from different sources that have been subject to a critical evaluation of their random and systematic errors.

Wilkins drew attention to the abstracts submitted by himself (36) and by Mrs Haramundanis (see end of section 3.1), and then put forward a view of how data centre operations might be organised for astronomy. He suggested that there should be several principal data-dissemination centres which would hold copies of the main datafiles and which would be prepared to provide copying and search services, and possibly facilities for visiting astronomers who wish to use the data interactively. They would cooperate with each other so as to share the tasks of merging catalogues, forming cross-identification indexes, etc. Each of these principal centres would serve a particular area and would, where necessary, make arrangements to obtain specialist data from one of the other centres. These centres should be in institutions in which astronomers are using the data for research purposes and providing new

data in their own fields. In addition there should be data-
analysis centres which would be concerned with the compilation of
catalogues of critically evaluated data in specialist fields;
they would not be expected to have extensive computing facilities,
and so they would normally pass their data to a data-dissemination
centre, and would not deal directly with requests for copies of
the files; they would, however, answer enquiries of a technical
character about the data, and perhaps supply new data that had
not yet been incorporated in the main file. Such an analysis
centre might even be located with a dissemination centre, or
have formal links with one, but the main aim would be to build
on the informal arrangements which have worked so well in astronomy in the past, and which have permitted small institutions to
make very valuable contributions to the astronomical database.

8. RECOMMENDATIONS ON IAU ACTIVITIES

During the final session there was an informal discussion
about the action that should be taken to implement some of the
suggestions for future activities that had been made during the
course of the colloquium. On the Chairman's suggestion it was
agreed that no attempt should be made to draft and pass formal
recommendations; rather, any such recommendations should be considered by IAU Commission 5 (Documentation) at Grenoble, during
the session concerned with the Working Group on Numerical Data.
Bidelman wondered whether the name of the Group is sufficiently
general to encompass the types of astronomical data that are not
strictly numerical, such as descriptions of spectroscopic peculiarities, and he suggested the name 'Working Group on Astronomical
Data'. Baum also considered that the name should avoid the infer-
ence that the group is interested only in conventional tabulations
of reduced data. Eichhorn suggested that the Group should become
a separate Commission but Wilkins drew attention to the considerable overlap with the other interests of Commission 5 and to the
fact that the decision to associate the Group with Commission had
been made only in 1973. Wood remarked that it would be easier
to persuade the Executive Committee to accept a change of name
than to establish a new commission.

There was general agreement to the Chairman's view that the
Group should prepare a guide to the presentation of astronomical
data in the primary literature.

The meeting then discussed Westerhout's suggestion that there
should be a group to keep up to date with hardware developments.
M.S. Davis spoke in support and suggested that the group could
also produce broad guidelines as to standard formats, choice of
data management systems, etc. Duncombe drew attention to the
financial constraints that limit the equipment available to data

centres, and said that when equipment is upgraded efforts should
be made to maintain compatibility so as to assure continued ease
of exchange between data centres. Westerhout thought that the
Group should report frequently and Jaschek offered to publish the
reports biannually in the Information Bulletin of the Strasbourg
Centre. The Chairman asked Westerhout and Davis to formulate
recommendations to only these lines.

The discussion of the action required to overcome the problems caused by the proliferation of designations for astronomical
objects led also to some discussion of technical matters that
have been reported in section 4. Jaschek considered that the best
approach is to convince the editors of leading journals that adequate identification should be given in every paper. Dixon drew
attention to the danger that a resolution on nomenclature may be
misunderstood; at the last IAU General Assembly, Commission 28
adopted a resolution that all future catalogues of galaxies, UV
objects, etc should be designated by Parkes-type numbers, eg
2024-18; unfortunately some authors misunderstood this and created new names for objects that already had other designations;
such action leads to confusion and provides no guide to readers
as to where to find the original catalogue. Bidelman favoured
the use of ordinary designations for stars brighter than the magnitude limit of the B.D. catalogue, since these would lead most
easily to further documentation. Eichhorn and Westerhout both
took the view that every object should have a precise, unambiguous
designation. It was generally agreed that the problem needed
detailed study by a small group, and that a new system could not
be introduced until the General Assembly in 1979. The Chairman
asked Bidelman and Eichhorn to prepare a recommendation on this
matter.

There was no dissent to Parsons' view that the bibliographic
files on stellar spectra and eclipsing binaries maintained by
Bidelman and Wood should be made more widely available in computer-
accessible form, nor to Lynga's suggestion that the Prague catalogue on stellar clusters should be similarly treated. There
were, however, divergent views on the suggestion by R.J. Davis
that an attempt should be made to compile on tape a comprehensive
catalogue of photographic-plate collections. Underhill felt that
the publication of such information should be left to the institution concerned. McCarthy commented that not all of the plates
would be worth examining. On the other hand, Westerhout considered
that a lot of valuable telescope time is wasted in duplicating
observations. Baum said that the listing of plate collections and
the dissemination of a merged list would be an exact analogue of
the 118 000-item listing of planetary photographs already compiled
and widely distributed by the Planetary Research Centre at the
Lowell Observatory; this planetary "catalogue" has substantially
increased the utilisation of the plate collections and has not

created any problems; he therefore did not agree with the reluc-
tance to encourage observatories to make plate listings freely
available and to merge them in due course. The matter was left
for further discussion within the Working Group.

9. CONCLUSIONS

Delhaye had to return unexpectedly to Paris and so was not
able to present his Concluding Remarks on the Colloquium.
Instead, Garstang (38) reviewed some of the ways in which data
could be made more readily available and intelligible to those
who will wish to find and use the data. Heintz (37) preceded his
vote of thanks to the organizers by discussing the role of the
abstracting services and of libraries in the dissemination of data.

The Chairman expressed his thanks to Jaschek and the other
staff at Strasbourg who had contributed to the organization of the
Colloquium; he thanked the participants for their contributions in
the presentation of papers and in the discussions; and he looked
forward to the continuation of the discussions on the improvement
of astronomical data services at the Grenoble meetings.

ASTROPHYSICS AND SPACE SCIENCE LIBRARY

Edited by

J. E. Blamont, R. L. F. Boyd, L. Goldberg, C. de Jager, Z. Kopal, G. H. Ludwig, R. Lüst,
B. M. McCormac, H. E. Newell, L. I. Sedov, Z. Švestka, and W. de Graaff

1. C. de Jager (ed.), *The Solar Spectrum, Proceedings of the Symposium held at the University of Utrecht, 26–31 August, 1963*. 1965, XIV + 417 pp.
2. J. Ortner and H. Maseland (eds.), *Introduction to Solar Terrestrial Relations, Proceedings of the Summer School in Space Physics held in Alpbach, Austria, July 15–August 10, 1963 and Organized by the European Preparatory Commission for Space Research*. 1965, IX + 506 pp.
3. C. C. Chang and S. S. Huang (eds.), *Proceedings of the Plasma Space Science Symposium, held at the Catholic University of America, Washington, D.C., June 11–14, 1963*. 1965, IX + 377 pp.
4. Zdeněk Kopal, *An Introduction to the Study of the Moon*. 1966, XII + 464 pp.
5. B. M. McCormac (ed.), *Radiation Trapped in the Earth's Magnetic Field. Proceedings of the Advanced Study Institute, held at the Chr. Michelsen Institute, Bergen, Norway, August 16–September 3, 1965*. 1966, XII + 901 pp.
6. A. B. Underhill, *The Early Type Stars*. 1966, XII + 282 pp.
7. Jean Kovalevsky, *Introduction to Celestial Mechanics*. 1967, VIII + 427 pp.
8. Zdeněk Kopal and Constantine L. Goudas (eds.), *Measure of the Moon. Proceedings of the 2nd International Conference on Selenodesy and Lunar Topography, held in the University of Manchester, England, May 30–June 4, 1966*. 1967, XVIII + 479 pp.
9. J. G. Emming (ed.), *Electromagnetic Radiation in Space. Proceedings of the 3rd ESRO Summer School in Space Physics, held in Alpbach, Austria, from 19 July to 13 August, 1965*. 1968, VIII + 307 pp.
10. R. L. Carovillano, John, F. McClay, and Henry R. Radoski (eds.), *Physics of the Magnetosphere, Based upon the Proceedings of the Conference held at Boston College, June 19–28, 1967*. 1968, X + 686 pp.
11. Syun-Ichi Akasofu, *Polar and Magnetospheric Substorms*. 1968, XVIII + 280 pp.
12. Peter M. Millman (ed.), *Meteorite Research. Proceedings of a Symposium on Meteorite Research, held in Vienna, Austria, 7–13 August, 1968*. 1969, XV + 941 pp.
13. Margherita Hack (ed.), *Mass Loss from Stars. Proceedings of the 2nd Trieste Colloquium on Astrophysics, 12–17 September, 1968*. 1969, XII + 345 pp.
14. N. D'Angelo (ed.), *Low-Frequency Waves and Irregularities in the Ionosphere. Proceedings of the 2nd ESRIN-ESLAB Symposium, held in Frascati, Italy, 23–27 September, 1968*. 1969, VII + 218 pp.
15. G. A. Partel (ed.), *Space Engineering. Proceedings of the 2nd International Conference on Space Engineering, held at the Fondazione Giorgio Cini, Isola di San Giorgio, Venice, Italy, May 7–10, 1969*. 1970, XI + 728 pp.
16. S. Fred Singer (ed.), *Manned Laboratories in Space. Second International Orbital Laboratory Symposium*. 1969, XIII + 133 pp.
17. B. M. McCormac (ed.), *Particles and Fields in the Magnetosphere. Symposium Organized by the Summer Advanced Study Institute, held at the University of California, Santa Barbara, Calif., August 4–15, 1969*. 1970, XI + 450 pp.
18. Jean-Claude Pecker, *Experimental Astronomy*. 1970, X + 105 pp.
19. V. Manno and D. E. Page (eds.), *Intercorrelated Satellite Observations related to Solar Events. Proceedings of the 3rd ESLAB/ESRIN Symposium held in Noordwijk, The Netherlands, September 16–19, 1969*. 1970, XVI + 627 pp.
20. L. Mansinha, D. E. Smylie, and A. E. Beck, *Earthquake Displacement Fields and the Rotation of the Earth, A NATO Advanced Study Institute Conference Organized by the Department of Geophysics, University of Western Ontario, London, Canada, June 22–28, 1969*. 1970, XI + 308 pp.
21. Jean-Claude Pecker, *Space Observatories*. 1970, XI + 120 pp.
22. L. N. Mavridis (ed.), *Structure and Evolution of the Galaxy. Proceedings of the NATO Advanced Study Institute, held in Athens, September 8–19, 1969*. 1971, VII + 312 pp.
23. A. Muller (ed.), *The Magellanic Clouds. A European Southern Observatory Presentation: Principal Prospects, Current Observational and Theoretical Approaches, and Prospects for Future Research, Based on the Symposium on the Magellanic Clouds, held in Santiago de Chile, March 1969, on the Occasion of the Dedication of the European Southern Observatory*. 1971, XII + 189 pp.

24. B. M. McCormac (ed.), *The Radiating Atmosphere. Proceedings of a Symposium Organized by the Summer Advanced Study Institute, held at Queen's University, Kingston, Ontario, August 3–14, 1970.* 1971, XI + 455 pp.
25. G. Fiocco (ed.), *Mesospheric Models and Related Experiments. Proceedings of the 4th ESRIN-ESLAB Symposium, held at Frascati, Italy, July 6–10, 1970.* 1971, VIII + 298 pp.
26. I. Atanasijević, *Selected Exercises in Galactic Astronomy.* 1971, XII + 144 pp.
27. C. J. Macris (ed.), *Physics of the Solar Corona. Proceedings of the NATO Advanced Study Institute on Physics of the Solar Corona, held at Cavouri-Vouliagmeni, Athens, Greece, 6–17 September 1970.* 1971, XII + 345 pp.
28. F. Delobeau, *The Environment of the Earth.* 1971, IX + 113 pp.
29. E. R. Dyer (general ed.), *Solar-Terrestrial Physics/1970. Proceedings of the International Symposium on Solar-Terrestrial Physics, held in Leningrad, U.S.S.R., 12–19 May 1970.* 1972, VIII + 938 pp.
30. V. Manno and J. Ring (eds.), *Infrared Detection Techniques for Space Research. Proceedings of the 5th ESLAB-ESRIN Symposium, held in Noordwijk, The Netherlands, June 8–11, 1971.* 1972, XII + 344 pp.
31. M. Lecar (ed.), *Gravitational N-Body Problem. Proceedings of IAU Colloquium No. 10, held in Cambridge, England, August 12–15, 1970.* 1972, XI + 441 pp.
32. B. M. McCormac (ed.), *Earth's Magnetospheric Processes. Proceedings of a Symposium Organized by the Summer Advanced Study Institute and Ninth ESRO Summer School, held in Cortina, Italy, August 30–September 10, 1971.* 1972, VIII + 417 pp.
33. Antonin Rükl, *Maps of Lunar Hemispheres.* 1972, V + 24 pp.
34. V. Kourganoff, *Introduction to the Physics of Stellar Interiors.* 1973, XI + 115 pp.
35. B. M. McCormac (ed.), *Physics and Chemistry of Upper Atmospheres. Proceedings of a Symposium Organized by the Summer Advanced Study Institute, held at the University of Orléans, France, July 31–August 11, 1972.* 1973, VIII + 389 pp.
36. J. D. Fernie (ed.), *Variable Stars in Globular Clusters and in Related Systems. Proceedings of the IAU Colloquium No. 21, held at the University of Toronto, Toronto, Canada, August 29–31, 1972.* 1973, IX + 234 pp.
37. R. J. L. Grard (ed.), *Photon and Particle Interaction with Surfaces in Space. Proceedings of the 6th ESLAB Symposium, held at Noordwijk, The Netherlands, 26–29 September, 1972.* 1973, XV + 577 pp.
38. Werner Israel (ed.), *Relativity, Astrophysics and Cosmology. Proceedings of the Summer School, held 14–26 August, 1972, at the BANFF Centre, BANFF, Alberta, Canada.* 1973, IX + 323 pp.
39. B. D. Tapley and V. Szebehely (eds.), *Recent Advances in Dynamical Astronomy. Proceedings of the NATO Advanced Study Institute in Dynamical Astronomy, held in Cortina d'Ampezzo, Italy, August 9–12, 1972.* 1973, XIII + 468 pp.
40. A. G. W. Cameron (ed.), *Cosmochemistry. Proceedings of the Symposium on Cosmochemistry, held at the Smithsonian Astrophysical Observatory, Cambridge, Mass., August 14–16, 1972.* 1973, X + 173 pp.
41. M. Golay, *Introduction to Astronomical Photometry.* 1974, IX + 364 pp.
42. D. E. Page (ed.), *Correlated Interplanetary and Magnetospheric Observations. Proceedings of the 7th ESLAB Symposium, held at Saulgau, W. Germany, 22–25 May, 1973.* 1974, XIV + 662 pp.
43. Riccardo Giacconi and Herbert Gursky (eds.), *X-Ray Astronomy.* 1974, X + 450 pp.
44. B. M. McCormac (ed.), *Magnetospheric Physics. Proceedings of the Advanced Summer Institute, held in Sheffield, U.K., August 1973.* 1974, VII + 399 pp.
45. C. B. Cosmovici (ed.), *Supernovae and Supernova Remnants. Proceedings of the International Conference on Supernovae, held in Lecce, Italy, May 7–11, 1973.* 1974, XVII + 387 pp.
46. A. P. Mitra, *Ionospheric Effects of Solar Flares.* 1974, XI + 294 pp.
47. S.-I. Akasofu, *Physics of Magnetospheric Substorms.* 1977, XVIII + 599 pp.
48. H. Gursky and R. Ruffini (eds.), *Neutron Stars, Black Holes and Binary X-Ray Sources.* 1975, XII + 441 pp.
49. Z. Švestka and P. Simon (eds.), *Catalog of Solar Particle Events 1955–1969. Prepared under the Auspices of Working Group 2 of the Inter-Union Commission on Solar-Terrestrial Physics.* 1975, IX + 428 pp.
50. Zdeněk Kopal and Robert W. Carder, *Mapping of the Moon.* 1974, VIII + 237 pp.
51. B. M. McCormac (ed.), *Atmospheres of Earth and the Planets. Proceedings of the Summer Advanced Study Institute, held at the University of Liège, Belgium, July 29–August 8, 1974.* 1975, VII + 454 pp.
52. V. Formisano (ed.), *The Magnetospheres of the Earth and Jupiter. Proceedings of the Neil Brice Memorial Symposium, held in Frascati, May 28–June 1, 1974.* 1975, XI + 485 pp.

53. R. Grant Athay, *The Solar Chromosphere and Corona: Quiet Sun.* 1976, XI + 504 pp.

54. C. de Jager and H. Nieuwenhuijzen (eds.), *Image Processing Techniques in Astronomy. Proceedings of a Conference, held in Utrecht on March 25–27, 1975*, XI + 418 pp.

55. N. C. Wickramasinghe and D. J. Morgan (eds.), *Solid State Astrophysics. Proceedings of a Symposium, held at the University College, Cardiff, Wales, 9–12 July 1974.* 1976, XII + 314 pp.

56. John Meaburn, *Detection and Spectrometry of Faint Light.* 1976, IX + 270 pp.

57. K. Knott and B. Battrick (eds.), *The Scientific Satellite Programme during the International Magnetospheric Study. Proceedings of the 10th ESLAB Symposium, held at Vienna, Austria, 10–13 June 1975.* 1976, XV + 464 pp.

58. B. M. McCormac (ed.), *Magnetospheric Particles and Fields. Proceedings of the Summer Advanced Study School, held in Graz, Austria, August 4–15, 1975.* 1976, VII + 331 pp.

59. B. S. P. Shen and M. Merker (eds.), *Spallation Nuclear Reactions and Their Applications.* 1976, VIII + 235 pp.

60. Walter S. Fitch (ed.), *Multiple Periodic Variable Stars. Proceedings of the International Astronomical Union Colloquium No. 29, Held at Budapest, Hungary, 1–5 September 1975.* 1976, XIV + 348 pp.

61. J. J. Burger, A. Pedersen, and B. Battrick (eds.), *Atmospheric Physics from Spacelab. Proceedings of the 11th ESLAB Symposium, Organized by the Space Science Department of the European Space Agency, held at Frascati, Italy, 11–14 May 1976.* 1976, XX + 409 pp.

62. J. Derral Mulholland (ed.), *Scientific Applications of Lunar Laser Ranging. Proceedings of a Symposium held in Austin, Tex., U.S.A., 8–10 June, 1976.* 1977, XVII + 302 pp.